podcasting

How to get started with podcasting in your organisation - second edition

social media
book one

Lee Hopkins

°138

Published by degrees138, Adelaide, South Australia—degrees138.com

I love deadlines. I love the whooshing noise they make as they go by.

— Douglas Adams

A common mistake that people make when trying to design something completely foolproof is to underestimate the ingenuity of complete fools.

— Douglas Adams

If you think you're too small to make an impact, try going to bed with a mosquito.

— Edith Sitwell

For my Agatha, who continues to amaze me with how much love is stored inside that petite body.

And to my Caz—despite being very senior in years, Caz is a Yellow Labrador who just keeps giving me unconditional love. I make every effort to give her unconditional love in return. A tummy rub and cuddle with her makes my day complete.

contents

about the author

Lee Hopkins is an Australian-based business communication consultant who helps clients develop communication strategies for an online environment. His clients include ANZ Bank; BT Financial Services; Bristol-Myers Squibb; Clarity Communications; Bower Place Consulting; Effective Consulting & Rehabilitation; Deakin University; International Association of Business Communicators, San Francisco; News Limited, London; University of South Australia; Public Relations Institute of Australia; RSPCA of SA; Supreme Education Council, Doha; TEDx Adelaide.

Lee is also a psychologist and has consulted as a management psychologist. He has a passion for employee communications and specialises in helping companies and fellow professionals discover the almost limitless possibilities for improved business communication that both 'old fashioned' and new technology offers.

A keen and active member of the International Association of Business Communicators (IABC), Lee is a widely read commentator on communication issues, and regularly showcases and demonstrates to businesses how they can 'marry' proven communication practices with new technology.

You can find a small list of Lee's clients here:

leehopkins.com/lee-hopkins-clients/

And a number of testimonials here:

leehopkins.com/lee-hopkins-testimonials/

about this book

A key component of social media, podcasts—be they audio or video—offer huge potential for organisations.

This book describes how you can get started with podcasting in your organisation. For those who've gone before you, it's safe to say that most organisations have overwhelmingly positive experiences with podcasting. From small to large organisations, there are many examples of podcast initiatives being piloted and going on to become raging successes.

Once you get over your initial fears, you'll see that podcasting is great fun and it's exciting, whether serious, light-hearted, or a mixture of both.

In addition to practical advice from myself and industry leaders on how to get started with podcasting, and suggestions for how they might be used at work, this mini-book also includes case study examples of podcasts in action.

what is podcasting?

ORIGINALLY DEFINED as an audio file that was delivered to you automatically via really simple syndication (RSS), the term "podcasting" has now come to embody many different facets of audio on the web.

A quick way of explaining them these days is to call them mini radio shows that you can listen to when and where you choose—be that at your computer during lunch, on the commute to work or walking your dog. You could be folding the laundry, painting the walls, or taking a road trip.

The content of that audio media varies considerably.

Quite often video is also bundled into the "podcast" category. Although this report doesn't go into detail about how to produce video for podcasts, or vidcasts as they're also known, I've included a couple of case studies to illustrate how some organisations are adding this to the mix.

The early adopters of this new information distribution channel called podcasting used it predominantly as their own music radio station; indeed, you could call podcasting 'radio with a rewind button'. But once the costs of production (almost zero) became known to businesses and consultants it was only a matter of time before all sorts of material became available by RSS subscription.

Today you can subscribe to:

- University lectures
- Comedy shows
- Music shows (focusing on each and every type of music—whatever your taste, you can find a niche specialist who equally shares your passion)
- Industry information
- Interest groups

For example, the Business Communication industry has many podcasts, most notably:

- **For Immediate Release** (forimmediaterelease.biz). Seasoned PR and Business Communication pros Shel Holtz (California) and Neville Hobson (England) with a podcast that stretches back to the very early days of social media and the birth of podcasting. Discussions range from communication fundamentals to leading edge developments and news. Essential listening for communicators.
- **Recipe for SEO Success** (therecipeforseosuccess.com/category/podcast/), one of the legendary Kate Toon's digital suite of services. The 'Recipe for SEO Success' course is both reassuringly expensive and over-subscribed twice a year, and there's always a waiting list to get on the course.
- Kate also runs the fabulous **Clever Copywriting School** (clevercopywritingschool.com/), and I am proud to be an alumni of both the Recipe course and the copywriting school.
- **Talked About Marketing**, the podcast run by 'Mr Podcast' Steve Davis and his business partner David Olney, with editing by Tim Whiffen (talkedaboutmarketing.com/podcast/).
- **The AI Train**, where your hosts, the crafty Content Strategist and Creative Copywriter Tim King, alongside the sharp-witted ChatGPT Trainer and Copywriter Leanne Shelton, stoke the boilers of insight and hurl coal into the fiery furnace of discourse on the wild, untamed frontier of Artificial Intelligence (shows.acast.com/the-ai-train). Yes, they wrote that. Good, isn't it?!

Anything that can be recorded and distributed as a digital audio file can be podcasted (remembering that the podcast isn't the audio file itself, but how it's delivered and distributed, which is via RSS, although thanks to technology RSS is now an invisible but still essential component).

when did podcasting begin?

PODCASTING, A PORTMANTEAU OF "IPOD" and "broadcasting", is a popular method for distributing digital audio content over the internet. It allows individuals or organisations to create and share their own audio shows or episodes, which can be accessed by listeners on various devices such as smartphones, tablets, and computers.

The beginning of podcasting can be traced back to the early 2000s, when former MTV video jockey Adam Curry and software developer Dave Winer were early adopters of the concept of "audio blogging". This involved uploading audio files onto a website or blog, which required listeners to manually download them in order to listen. I think I was the first podcaster, however, because in 1996 I used to record then publish a daily news service mp3 (maximum five minutes long) on topics of interest for small businesses, which they could download and listen to at their convenience, or play on a NewsCorp-owned website.

I started doing this on a whim and didn't see anyone else doing it, so it was a differentiator for the visitors to one of Rupert Murdoch's web businesses in London that enabled him to publish his newspapers' content on the web .

However, with the release of Apple's iPod in 2001 and the subsequent rise of portable digital media players, the potential for on-the-go audio consumption began to emerge.

In 2003, Curry and Winer collaborated to develop a new method for distributing audio content that would automatically download new episodes onto portable devices. This led to the creation of RSS (Really Simple Syndication = a standardized system for the distribution of content from an online publisher that others can subscribe to) feed enclosures, which allowed users to subscribe to and receive updates from their favourite podcasts through an automated process. This development marked the true beginnings of podcasting as we know it today.

The term "podcast" was first coined by journalist Ben Hammersley in an article for *The Guardian* newspaper in 2004, and since then has become a widely recognised term for this form of digital audio content distribution. That same year, former MTV video jockey and radio host Christopher Lydon launched the first podcast radio show, "Radio Open Source". This groundbreaking move paved the way for other radio stations and broadcasters to embrace podcasting as a new platform for content dissemination.

Podcasting continued to gain popularity in the mid-2000s with the rise of platforms such as iTunes, which made it easier for listeners to discover and subscribe to podcasts. In 2005, Apple introduced the first version of iTunes with built-in support for podcasts, and just a year later millions of people were using the platform to download and listen to their favourite shows.

One key factor in podcasting's early success was its accessibility and low barrier to entry. With relatively simple equipment and software, anyone could create and publish their own audio content without needing a radio station or broadcasting license. This democratisation of audio content creation allowed for a diverse range of voices and topics to be heard, making podcasting a truly democratic medium.

As podcasting continued to grow in popularity, more and more businesses and organisations began to see its potential for marketing and advertising. By 2007, McDonald's had launched its first branded podcast, "The Feed", which featured discussions on pop culture and music. This paved the way for other companies to follow suit, utilising podcasting as a platform for brand awareness and reaching new audiences.

Today, podcasting has become a thriving industry with an estimated 700,000 active podcasts and over 29 million episodes available worldwide. It continues to evolve and innovate, with advancements in technology making it easier for creators to produce high-quality audio content and for listeners to access and interact with their favourite shows.

From its humble beginnings as an experimental form of online radio, podcasting has become a powerful medium for storytelling, education, entertainment, and more. It has given a voice to individuals and communities around the world, connecting people through shared interests and passions. And with its continued growth and advancements, the future of podcasting looks bright. So, it can be said that the beginnings of podcasting have not only shaped its present but also paved the way for its promising future.

For example, we are already seeing a rise in live podcasts and interactive content, as well as increased investment from major media companies such as Spotify and Amazon.

For your marketing team, the question is not 'should we?' but 'when will we?'

why is podcasting so popular?

PODCASTING HAS BECOME INCREASINGLY popular in recent years, and for good reason. It offers a unique and convenient way to consume content that appeals to a wide range of interests. Here are some reasons why podcasting has captured the attention of millions:

- **Convenience**: With busy schedules and on-the-go lifestyles, many people find it difficult to make time for reading or watching video content. Podcasting allows listeners to consume content while multitasking, whether it's during their commute, workout, or daily chores.

- **Diverse topics**: From education and news to entertainment and self-improvement, there is a podcast for every interest. With so many different genres and niche subjects available, listeners can find something that caters specifically to their interests.

- **Authenticity**: Podcasts often feature real people having authentic conversations, making the content more relatable and engaging for listeners. This human touch sets podcasts apart from other forms of media that may feel more scripted or polished.

- **On-demand access**: Unlike traditional radio or television shows, podcasts offer on-demand access to their episodes. Listeners can choose when and where they want to listen, rather than being tied to a specific broadcast schedule.

• **Diverse voices and perspectives**: Podcasting has given a platform to diverse voices and perspectives that may not have been represented in mainstream media. This allows for a more inclusive and well-rounded exchange of ideas and information.

• **Low barrier to entry**: Starting a podcast does not require expensive equipment or a large production team. With basic recording equipment and some editing software, anyone can create their own podcast and share their content with the world.

• **Engaging storytelling**: Many podcasts use creative storytelling techniques to captivate listeners and keep them coming back for more. This can include sound effects, music, and expertly crafted narratives that draw listeners in and hold their attention.

• **Community building**: Podcasts often have a dedicated and loyal fan base, creating a sense of community amongst listeners. This can foster discussions and connections among like-minded individuals who share a passion for the same podcast topics.

Overall, the unique combination of convenience, diverse content, authenticity, and community building has made podcasting a popular form of media that shows no signs of slowing down. With new podcasts constantly emerging on a variety of topics, the future of podcasting looks bright. So, it's no surprise that more and more people are turning to this medium for their entertainment and information needs. So if you haven't already jumped on the podcast bandwagon, now is the perfect time to start exploring all that this popular form of media has to offer. Who knows, you may just find your new favourite podcast! Happy listening!

So, what are you waiting for? Grab your headphones and start exploring the world of podcasting today. With so many options to choose from, there's something for everyone. Whether you're looking for educational content, entertainment, or just a good laugh, podcasts have got you covered. And with the convenience of on-demand access, you can listen whenever and wherever it's convenient for you. So don't miss out on this popular and ever-growing medium—start discovering new podcasts now! Keep listening and stay curious. The possibilities with podcasting are endless.

what sort of business things can you podcast?

HERE ARE JUST A FEW EXAMPLES:

- Meetings
- Executive speeches
- Business updates
- Investor road shows (so employees can have some insight into what outside audiences are hearing)
- Customer trade show presentations (ditto above)
- Presentations by sales executives (so employees can hear how the company's products and services are sold to the outside world)

I even see different business units offering their own podcasts, allowing employees to subscribe to those that would be useful and ignore those that aren't. The IT department can hard-code those podcasts deemed important to everyone into their smartphone, just as they can make the intranet homepage the browser's default homepage. Access to external podcasts could easily be restricted, protecting the company's precious and costly bandwidth.

As many commentators are now saying, considering the low-to-no-cost entry into podcasting, its acceptance into the internal communication mix should happen sooner rather than later.

. . .

Save your busy executive from the tedium of countless non-productive hours. Here's how...

Let's consider the average manager or executive.

They receive 60-70 e-mails a day, links to spreadsheets and pages on the company intranet, links to other websites. Then they take a trip to another city. Hop on a plane... but now what?

If they're in "cattle class" (economy class), there's not enough room to swing a cat, let alone open their laptop.

They could probably read some of the material their colleagues sent them, but there's no printer available so unless they printed off the equivalent of *War and Peace* before they left, so they're stuck with the airline magazine.

But if they brought their phone preloaded with some episodes of their favourite podcasts they can listen to them. And also on their phone they could listen to:

- A recent presentation the CEO gave to the investment community.
- An update on sales figures from an affiliate organisation.
- A synopsis of performance by the executive's own division.
- News and commentary from leading online broadcasters relevant to the executive's own industry.
- A few really banging tunes to help the executive take their mind off their troubles and stress!

how can i get great content for my podcast?

ANOTHER QUESTION I'm often asked is "how can I get great content without having to hire a professional radio announcer (or pretend to be one)?" Here are a few simple examples:

- Ask the sound guy at the CEO's presentation to save the speech as a digital file.
- Adam in Finance can read the sales figures into a cheap microphone attached to his computer. A few clicks of his mouse and he's saved the file as an MP3. Suddenly Adam is a regular contributor and expert, with his own segment.
- Rob in Admin can read the performance stats into a cheap microphone attached to his computer with a few clicks of his mouse. Suddenly Rob is a regular contributor and expert, with his own segment.

Once you've decided to start a podcast, it's important to have great content to keep your listeners engaged and coming back for more. Here are some tips on how to get great content for your podcast:

1 **Know your audience**: The first step in creating great content is knowing who you're creating it for. Think about your target audience—what topics would interest them? What questions do they have? Under-

standing your audience will help you create content that resonates with them.

2 **Research and plan**: Take some time to research topics related to your podcast's theme. Look for trends, current events and interesting stories that you can talk about on your show. Make a plan for each episode, including an outline of the main points you want to cover and any guests you want to invite.

3 **Be authentic**: The best podcasts are those that feel genuine and authentic. Don't try to be someone you're not—let your personality shine through in your content. This will help you connect with your audience and keep them engaged.

4 **Invite guests**: Inviting guests on your podcast can bring fresh perspectives and add variety to your content. Look for experts in your podcast's niche or people with interesting stories to share. Not only will this provide great content, but it can also help you reach a new audience.

5 **Get feedback**: Don't be afraid to ask for feedback from your listeners. This can help you understand what resonates with them and what they want more of on your podcast. You can also use this feedback to improve your content and keep your audience interested.

6 **Stay consistent**: Consistency is key when it comes to creating great content for your podcast. Stick to a regular schedule so that your listeners know when to expect new episodes. This will help you build a loyal audience who will keep coming back for more.

In addition, here are some general ideas to consider when creating content for your podcast:

• Use storytelling techniques to engage your audience and make your content more interesting.

• Incorporate different formats, such as interviews, solo episodes, or panel discussions, to add variety to your show.

• Utilise social media platforms to gather topic suggestions from your listeners and involve them in the conversation.

• Collaborate with other podcasters to exchange ideas and potentially collaborate on episodes.

• Keep up with current events and trends in your niche, as these can provide great inspiration for content.

Remember, the most important thing is to be passionate about what you're talking about. This will come through in your content and keep your listeners interested. With these tips in mind, you'll be well on your way to creating great content for your podcast.

Also, don't be afraid to step out of your comfort zone and try new ideas and topics. Your listeners will appreciate the effort and engagement you bring to your podcast. Keep pushing yourself to create unique, informative, and entertaining content that will keep your audience coming back for more. And most importantly, have fun with it! After all, that's why you started a podcast in the first place. Keep experimenting, learning, and growing as a podcaster, and your content will continue to evolve and improve. Happy creating!

And don't forget to promote your podcast! Utilise social media, collaborate with other podcasters, and reach out to potential guests or sponsors. The more exposure you get, the more opportunities for great content will come your way. Keep engaging with your audience and incorporating their feedback into your content—after all, they are the reason you're creating this podcast in the first place. With dedication, hard work, and a passion for your topic, you'll continue to produce great content that will keep your listeners hooked and coming back for more.

how do i get the podcast delivered to the people i want to listen to it?

THERE ARE several ways to get your podcast delivered to the people you want to listen to it. One way is by submitting it to popular podcast directories such as Apple Podcasts, Google Podcasts, Spotify, and Stitcher. These directories allow listeners to easily find and subscribe to your podcast.

Another way is by promoting your podcast on social media platforms like Twitter, Facebook, and Instagram. This can help attract new listeners and also allow you to engage with your existing audience.

Additionally, you can create a website for your podcast where listeners can access all the episodes and get updates on upcoming episodes. This also gives you the opportunity to interact with your audience through comments and feedback.

Collaborating with other podcasts or influencers in your niche is also a great way to reach a larger audience. This can be done through guest appearances on other podcasts or cross-promoting each other's shows.

Lastly, word-of-mouth is still a powerful tool for promoting your podcast. Encourage your listeners to share and recommend your podcast to their friends and family.

Remember, the key to getting your podcast delivered to the people you want is by consistently producing high-quality content and actively promoting it through various channels. So, keep creating great episodes and don't be afraid to try out new ways to reach your target audience.

Just like any other form of content creation, consistency is key when it comes to delivering your podcast to the right people. This means sticking to a regular publishing schedule and continuously engaging with your listeners through social media and other platforms. It's important to also listen to feedback from your audience and make adjustments accordingly to keep them engaged.

Another effective way to get your podcast delivered is by utilising email marketing. By building an email list, you can directly reach out to your subscribers with updates and new episodes. This can be a great way to nurture a dedicated fan base and promote your podcast to a targeted audience.

Furthermore, consider collaborating with other podcasts or hosting live events to expand your reach. This can create buzz around your show and attract new listeners.

Lastly, don't forget the power of SEO (search engine optimisation). By optimising your podcast title, description, and keywords, you can increase the chances of it appearing in search results and reaching a wider audience.

In conclusion, there are various ways to get your podcast delivered to the people you want to listen to it. By utilising a combination of different strategies, consistently creating high-quality content, and actively promoting your show, you can effectively grow your audience and reach your target listeners. Keep experimenting and finding new ways to connect with your audience, and your podcast will continue to thrive.

what sort of software do i need?

CREATING a podcast may seem like a daunting task, but with the right software, it can be easily achievable. The type of software you need will depend on your level of experience and the complexity of your podcast.

If you are new to podcasting and want a simple, user-friendly option, there are many online platforms such as Anchor or Buzzsprout that offer all-in-one solutions for recording, editing, and hosting your podcast. These platforms often have templates and easy-to-use tools to help you create a professional-sounding podcast without any prior knowledge.

For those with more experience or looking for more control over their podcast, there are various software options available for recording and editing. Some popular choices include Audacity, GarageBand (for Mac users), or Adobe Audition. These software options often have more advanced features and allow for a higher level of customisation.

In addition to recording and editing software, you may also need audio equipment such as a microphone, headphones, and a pop filter to ensure high-quality sound for your podcast. However, if you are just starting out, most laptops and smartphones have built-in microphones that can produce decent sound quality.

It's important to note that the software you choose will also depend on your budget. While some options may be free or have a one-time purchase fee, others may require a monthly subscription. It's always recommended to do research and read reviews before committing to a particular software.

Lastly, if you plan on having guests on your podcast remotely, you may also need software for recording and managing remote interviews. Some options for this include Zencastr, Squadcast, or Skype.

Overall, the type of software you need to create a podcast will vary depending on your personal preferences and technical abilities. It's important to consider your goals and budget when choosing the right software for your podcasting journey. With the wide range of options available, there is something for everyone to create and share their podcast with the world. So don't be afraid to explore, experiment, and find the perfect software that works for you.

Some additional tips:

• Make sure to regularly update your software to ensure optimal performance.

• Don't be afraid to reach out for help or tutorials if you are struggling with a particular software. YouTube has a zillion how-to videos on typical software challenges and problems.

• Consider investing in some soundproofing materials or recording in a quiet environment to improve the overall quality of your podcast.

• Have fun with it and let your creativity shine! Your podcast is a reflection of you, so make it unique and enjoyable for both yourself and your listeners.

Keep learning and improving, and don't be afraid to try out new software or techniques to enhance your podcasting experience. The possibilities are endless in this ever-evolving medium.

Create, share, and inspire others through the power of podcasting. Keep creating, keep learning, and keep sharing your unique voice with the world! Keep exploring new software and techniques to take your podcast to the next level. The possibilities are endless in this ever-evolving medium. So don't be afraid to try new things and let your creativity shine.

Remember, the software is just a tool to help bring your podcast to life. It's up to you to provide engaging content and connect with your audience. So keep creating and sharing your unique voice with the world, and who knows where your podcasting journey will take you.

Don't let the fear of technology hold you back from creating a successful podcast. Keep exploring new software and techniques to enhance your podcasting experience. The more you experiment and learn, the better your podcast will become. And with a supportive community of fellow podcasters, you'll always have someone to turn to for advice and inspiration.

what equipment do i need?

"WHAT'S **the best equipment to buy for podcasting?"**

Podcasting is a popular and growing medium for sharing information, stories and entertainment. It allows people to create audio content and share it with the world. With the rise of smartphones and easy access to recording equipment, anyone can start their own podcast. But what tools do you need to create a successful podcast? ll discuss some essential tools that every podcaster should have.

1. Microphone

The most important tool for creating a podcast is a good quality microphone. A microphone will help you capture clean and clear audio, which is crucial for keeping your listeners engaged. There are many types of microphones available such as USB microphones, dynamic microphones, and condenser microphones. It's essential to do some research and choose the type that best suits your needs and budget.

2. Audio editing software

Once you have recorded your podcast, you'll need to edit it before publishing it. This is where audio editing software comes in. There are many free and paid options available such as Audacity, GarageBand, Adobe Audition and Hindenburg Journalist Pro. These software allow you to edit, mix and enhance your audio recordings, making them sound more professional.

3. Headphones

Headphones are an essential tool for podcasters as they allow you to monitor the audio quality while recording and editing. It's important to invest in a good pair of headphones that provide accurate sound reproduction. Closed-back headphones are preferable as they help block out external noise and prevent sound leakage.

4. Recording device

If you're recording on-the-go, a portable recording device can come in handy. These devices allow you to record high-quality audio while on the move. Some popular options include Zoom H6 Handy Recorder and Tascam DR-40X 4-Track Audio Recorder. They offer multiple inputs for microphones, making it easier to record interviews and discussions with multiple people.

5. Pop filter

A pop filter is a small but essential tool that helps reduce or eliminate harsh plosive sounds, such as "p" and "b," when recording vocals. It's a simple yet effective way to improve the audio quality of your podcast recordings. You can either buy a physical pop filter or use a DIY alternative such as pantyhose or tights stretched over a thin metal coat-hanger (like the ones you get from your dry cleaner) that's been bent to a suitable shape.

6. Hosting and Distribution platform

Once you have created your podcast, you'll need to host it somewhere so that people can listen to it. Some popular hosting and distribution platforms include Podbean, Buzzsprout, and my personal favourite since the dawn of podcasting: Libsyn. These platforms allow you to upload your audio files and distribute them to major podcast directories such as Apple Podcasts and Spotify.

7. Marketing and Promotion tools

Creating a great podcast is only half the battle; you also need to market and promote it to reach a wider audience. There are various tools available to help with this, such as social media platforms like Twitter and Instagram, email marketing services like Mailchimp, and podcast directories like Podcast Addict. It's essential to use these tools to your advantage and engage with your listeners regularly.

In conclusion, creating a podcast requires some essential tools, but it doesn't have to break the bank. With advancements in technology, there are many affordable options available for aspiring podcasters. Invest in good quality equipment and be creative with using free or DIY alternatives when possible. With the right tools and dedication, you can create a successful podcast that will captivate your audience and keep them coming back for more. So what are you waiting for?

Keep exploring, learning, and experimenting with different tools to take your podcast to the next level. As technology continues to advance, there will always be new and innovative tools available to enhance the podcasting experience. Don't be afraid to try out new equipment and techniques, and don't forget to constantly engage with your listeners for valuable feedback.

Remember, the key to a successful podcast is not just having the right tools but also using them effectively. So take your time, do your research, and choose the best tools that suit your needs and budget. With hard work, dedication, and the right tools, you can create a podcast that will stand out in the crowded world of podcasting.

who is podcasting?

THERE ARE many successful businesses that have incorporated podcasting into their overall marketing strategy. One of the most notable examples is Spotify, a music streaming service that has expanded to include podcasts in recent years. In fact, Spotify now produces its own original podcasts in addition to hosting popular shows from other creators.

Another business that has seen success with podcasting is Mailchimp, an email marketing service. They launched their first podcast, "Going Through It," in 2018 and have since produced several other podcasts that focus on different topics such as entrepreneurship and small business growth.

Other successful businesses that have embraced podcasting include TED Talks, a nonprofit organisation that produces educational talks on a wide range of topics, and Shopify, an ecommerce platform for online businesses. Both of these companies have created podcasts that align with their brand and provide valuable content to their audience.

In addition, many media companies such as The New York Times, CNN, and BBC also have successful podcasting ventures. These businesses use podcasts as another platform to share news and information with their audiences, reaching a wider demographic than traditional print or broadcast mediums.

Overall, it is clear that podcasting has become a valuable tool for businesses to connect with their audience, establish thought leadership, and promote their brand. As the popularity of podcasts continues to grow, we can expect to see even more successful businesses incorporating this medium into their marketing strategies. So, if you're considering starting a business or looking to expand your existing one, it may be worth considering adding podcasting to your marketing mix. Not only can it help you reach a wider audience, but it also allows for more personal and authentic connections with your customers.

Podcasting has also proven to be a valuable tool for businesses looking to increase their online presence and improve their search engine rankings. By regularly producing high-quality podcast episodes, businesses can attract more listeners and increase website traffic. Furthermore, podcasts can also be repurposed into written content for blogs or social media posts, providing even more opportunities to reach potential customers.

The success of businesses in using podcasting as a marketing strategy has also led to the rise of podcast production companies. These companies offer services such as recording, editing, and distributing podcasts for businesses that may not have the resources or expertise to produce their own shows. This further highlights the importance of podcasting in today's business world.

Moreover, podcasting allows businesses to connect with their audience on a deeper level by providing valuable insights, advice, and information related to their industry. This not only builds trust and credibility with customers but also positions the company as an expert in their field.

In conclusion, podcasting has become an essential component of successful businesses today. Whether you are a small startup or a well-established company, incorporating podcasting into your marketing strategy can greatly benefit your business and help you stand out in a crowded market. So why not join the podcasting trend and see how it can elevate your business to new heights? So go ahead, start recording, and let your voice be heard in the world of podcasting. The potential for growth and success is limitless.

. . .

These are just some of the organisations both large and small that are podcasting:

- ABC Australia
- Adobe
- The Age
- Amazon
- Auburn University
- Australian Flexible Learning Framework
- Australian National University (ANU)
- Bloomberg
- BMW
- Computerworld
- Cisco
- Deloitte
- Disneyland
- Gartner
- General Motors
- Georgetown University
- Harvard Business School
- Hewlett Packard
- IBM
- iPressRoom
- Jupiter Research
- McDonalds
- Mercedez Benz
- NASA
- National Oceanic & Atmospheric Administration (NOAA), US Department of Commerce
- New Jersey Bankers Association
- Novell
- Oracle
- Ottawa Centre for Research and Innovation
- Peugeot
- Princeton
- Purina

- Renault F1 team
- Rutgers University
- San Francisco Museum of Modern Art
- SAP
- Simon & Schuster
- SkyNews
- Sprint
- St John Ambulance (UK)
- University of California
- Virgin Atlantic
- Whirlpool
- Williams Law Firm

other great reasons to podcast internally

REMEMBER way back when companies used to keep their sales forces "skilled up" via audio cassettes? The sales force used to listen to the cassettes in their car (on the way to work, home, client visits, etc.). Sales people had the option to listen to the communications at the time and location of their choice, all the while not interfering with their actual sales activities.

The cassettes took time to write and produce, sure, plus you needed to add on duplication and distribution costs and timeframes, but for the dissemination of non-urgent material it was a brilliant solution.

I've listened to countless hours of audio material from personal improvement publishers Nightingale-Conant and others over the years in my own "mobile university" (aka my car).

Which brings me back to podcasting.

Podcasting is a brilliant, low-cost way of getting mildly urgent and non-urgent communication out to relevant interest groups. What makes podcasting so cool is that the files can be downloaded to your smart-phone or computer and listened to while walking the dog, sitting on the bus or train, driving the car, sitting under a tree, enduring long and tedious flights, waiting in airport terminals, switching out taxi drivers who could run the country better than the current mob if only given a chance, and so on.

A podcast can be quickly and automatically downloaded onto a smartphone, ready for others to listen to on their way to work in the morning. How cool and efficient is that?

HOW DO I CREATE A PODCAST?

Simple—record someone or something, edit it and then publish it.

Record: using a microphone (a $20 version works just as well as a $100 version for most podcasts), record an interview, yourself or someone else.

Edit: take out the "ums and ahs", the fluffs, the mispronunciations, the nervous ticks (I have a dreadful habit of sucking in air through my front teeth when breathing between sentences, which I have to laboriously edit out!) and generally tighten up your recording. Software such as the free and fully-featured **Audacity**, the not-free **Adobe Audition**, and also the with-cost **Sound Forge Pro** allow you to add "compression" and change equalizer settings to make the sound a little more polished, but don't worry about those bells and whistles at first; all of that stuff can come later as you gain more experience and confidence. Add in some music (if your podcast format has music in it as introductory or outgoing "signature" pieces, for example) and sound effects if you want. Listen to the final output several times—you'd be amazed at how the recording you listened to at 4pm and thought was rather good sounds less so at 9.30 the next morning!

Publish: depending on your IT department's views, you may be able to host your podcast on the company server, or you may need to host it on a third-party's server (I personally recommend **Libsyn** for cost effectiveness and reliability). Once it has been hosted, make sure you let everyone in your world know about it—customers, employees, suppliers—depending, of course, on who your target audience is.

And make sure that the tech-savvy among your audience are able to subscribe to your podcast, "easy" when your users subscribe to your podcast if you give them the link to your **iTunes** or **Google Play** software, or when you host with **Libsyn**.

today's podcast landscape for business communicators and pr professionals

IN THE BOUNDLESS world of marketing and PR, certain industry leaders stand out for their innovative approaches and unique insights, many of which they share on their respective podcasts. These podcasts, available on various platforms like Spotify, Apple Podcasts, and Google Podcasts, serve as invaluable resources for professionals and enthusiasts alike.

One such leader is Seth Godin, a marketing guru known for his incisive perspectives on branding. His podcast, "Akimbo," is a treasure trove of insights, where he discusses the intersection of culture, marketing, and business. Following Seth closely is Gary Vaynerchuk, who has made a name for himself as a marketing visionary. His podcast "The GaryVee Audio Experience" provides a mix of his popular #AskGaryVee show, interviews, and keynote speeches on marketing and entrepreneurship.

The realm of Public Relations is dominated by players who wield their expertise to shape the perception of brands and individuals. One such titan is Gini Dietrich, the founder and CEO of Arment Dietrich. Her podcast "Spin Sucks" offers actionable advice on communication and PR. Ketchum PR's Barri Rafferty hosts "The Business of Story," a podcast that leverages the power of storytelling to create compelling PR narratives.

Next up, we can't ignore the impact of HubSpot in the digital marketing realm. The "HubSpot Academy Podcast" provides listeners with lessons and insights on sales, marketing, and customer service. For social media fanatics, Michael Stelzner's "Social Media Marketing Podcast" by Social Media Examiner is a must-listen. It offers new strategies and advice on social media.

Neil Patel and Eric Siu, renowned figures in the digital marketing sphere, host the "Marketing School" podcast. Here, they provide actionable marketing tips and strategies in bite-sized episodes. For those more interested in content marketing, Joe Pulizzi and Robert Rose offer "This Old Marketing Podcast," where they discuss content marketing trends and offer advice on strategy.

In the PR podcast landscape, we have "PR Week," hosted by PRWeek editors. This podcast stands as a beacon for PR professionals, featuring interviews with PR and communications leaders. Similarly, "For Immediate Release" by Shel Holtz is a bi-weekly podcast that covers contemporary communication trends and implications.

Steve Davis and David Olney, both marketers, produce the very successful and insightful marketing podcast, 'TAM: The Talked About Marketing podcast'. Each podcast episode tips its hat to Philip Kotler's famous "4 Ps of Marketing" (Product, Price, Place, Promotion), by honouring Davis and Olney's own 4 Ps of Podcasting: Person, Principles, Problems, and Perspicacity.

These are only a few of the podcasts hosted by the doyens of the marketing and PR industries. Each offers a unique perspective, allowing listeners to glean insights, discover trends, and stay abreast of industry developments. So, whether you're a marketing whiz or a PR pro, make sure to give these podcasts a listen and soak up the wisdom of the leaders in these fields.

how to ensure your podcast's success in 2024: a comprehensive guide

IN RECENT YEARS, podcasting has exploded in popularity as a medium for entertainment, education, and information. According to Statista, there were over 1.75 million active podcasts and more than 43 million episodes available worldwide as of October 2021 (statista.com/statistics/1201832/number-podcasts-worldwide/) and the numbers are expected to continue growing in the coming years.

With such a crowded podcast market, how can you ensure that your podcast stands out and remains successful in the year 2024? In this comprehensive guide, we will explore key strategies and tips for creating and maintaining a successful podcast that will thrive in the ever-evolving world of podcasting.

1. Identify your niche and target audience

The first step to creating a successful podcast is to identify your niche and target audience. With so many podcasts available, it's important to differentiate yourself by focusing on a specific topic or theme that appeals to a particular audience. This could be anything from true crime to personal development, from sports to finance.

Once you have identified your niche, it's important to understand your target audience. Who are they? What are their interests and preferences?

Knowing your target audience will help you tailor your content and marketing efforts to better reach and engage them.

2. Create high-quality content

The success of any podcast ultimately depends on the quality of its content. It's vital to consistently produce engaging, informative, and well-produced episodes that keep your audience coming back for more.

One way to ensure high-quality content is by creating a structure for your episodes. This could include an introduction, main topic, guest interviews or discussions, and a conclusion. Make sure to also do thorough research on each episode's topic and have a clear and organised outline before recording.

Moreover, don't be afraid to experiment with different formats such as solo episodes, interviews, or panel discussions. This will keep your podcast fresh and interesting for your listeners.

3. Utilise Social Media and other platforms

In today's digital age, having a strong online presence is crucial for any business or brand, including podcasts. Utilise social media platforms such as Twitter, Instagram, and Facebook to promote your podcast and engage with your audience.

In addition, consider utilising other platforms such as YouTube or live streaming on platforms like Twitch to reach a wider audience and provide additional content for your listeners.

4. Consistency is key

Consistency is key when it comes to maintaining a successful podcast. This means consistently producing new episodes on a regular schedule and maintaining a consistent tone, format, and quality in your content.

Having a set schedule for releasing episodes will also help build anticipation and loyalty from your audience. Make sure to communicate any changes or breaks in your schedule with your listeners to keep them informed and engaged.

5. Engage with your audience

Engaging with your audience is crucial for building a loyal and dedicated following. Encourage your listeners to give feedback, ask questions, or suggest topics for future episodes.

Responding to comments and messages on social media is also important in fostering a connection with your audience. You can even consider creating a community for your listeners to connect with each other and discuss the podcast's content.

6. Collaborate with other podcasters

Collaborating with other podcasters can be beneficial for both parties as it exposes each other's audiences to new content and potentially offers new perspectives or expertise on topics. Look for opportunities to guest host on other podcasts or invite guests onto your own show.

This can also help broaden your reach and attract new listeners who may not have otherwise discovered your podcast.

7. Continuously improve and adapt

The world of podcasting is constantly evolving, and it's important to continuously improve and adapt in order to stay relevant and successful. Stay up-to-date with industry trends, listen to feedback from your audience, and be open to trying new ideas or formats.

Regularly evaluating the success of your podcast and making necessary adjustments will help you stay ahead of the game and continue to attract and retain listeners.

. . .

Conclusion

Creating a successful podcast in 2024 requires dedication, hard work, and a willingness to continuously improve. By identifying your niche and target audience, consistently producing high-quality content, utilising social media and other platforms, maintaining consistency, engaging with your audience, collaborating with other podcasters, and adapting to the ever-changing industry, you can ensure your podcast's success in the years to come. So start planning, recording, and promoting your podcast now and get ready for a successful future in podcasting.

the five golden rules of podcasting

I believe there are five golden rules you need to know about if you are keen to create a great podcast:

1. **Understand the media.** You do this by downloading and listening to lots of podcasts from all different genres—comedy, other business podcasts, music, chat, special interests and so on. Listen to the podcasts, see what you like in each one and what you don't like, find out if there are some key elements that work for you, and you believe you can incorporate into your own podcast. Take lots of notes.
2. **Create a script.** Or at least a set of bullet points for yourself or the presenter, so that you/they don't wander too far off the track when the microphone is on.
3. **Use a template.** By that I mean decide what your format will be (light- hearted articles read by the one person, a two or three person chat, special-interest focused recordings, etc.) and what standard items there will be in each podcast edition (for example, the same intro/outro music, the same way of introducing the podcast edition, the "layout" of the podcast—e.g. intro theme, welcome, item 1, item 2, comments, closing theme).

4. **Edit and polish.** You can always do a number of "takes" when recording, then edit out the ums, ahhhs, fumbles and mispronunciations later. You can also edit out over-long pauses to liven up an otherwise slow and boring interviewee. Don't go overboard on your edits—you want to keep a natural feeling to your podcast and keep it from sounding too over- produced and slick—but equally don't ruin great content by not wielding the editing mouse firmly but fairly.
5. **Publish it.** Create a publishing schedule for each edition so that listeners will come to unconsciously expect your podcast to appear on their smartphone or on their computer. And tell the world about your new podcast. Splash it all over your intranet. If it's for the "outside world" to hear, then make sure every person in the company who has any interactions with customers, clients or people outside the company lets them know about the new podcast and how they can easily subscribe, download and listen to it. Email signatures are great for this—pop a link to the podcast in the email signature, and even a short line or two on why the email recipient should listen.

what social media can do for podcasting

THERE ARE things that social media can do, best expressed by the term RRISC:

- **Reputation** – Social media can help a company bolster its reputation, both online and off;
- **Rumours** – Social media can correct inaccurate rumours;
- **Information** – Social media can provide correct information;
- **Story** – Social media can help tell a company's story in a compelling manner; and
- **Customers** – Social media can address unhappy customers and help turn them into happy ones. If this is done publicly, you can bet the majority of the audience are watching closely to see how you deal with unhappy customers, helping inform their opinion of your organisation, and helping them decide whether they want to continue to do business with you or not.

With the rise of social media and its impact on various industries, it's no surprise that podcasting has also been greatly influenced by these platforms. Social media has become an essential tool for podcasters to promote their shows, engage with their audience, and grow their following.

One of the biggest advantages of using social media for podcasting is the ability to reach a wider audience. With millions of active users on platforms like Facebook, Instagram, and Twitter, podcasters can tap into a much larger pool of potential listeners. This not only helps in increasing the visibility of a podcast but also allows for easy sharing of episodes with friends and followers.

Moreover, social media provides an interactive platform for podcasters to engage with their audience. Through comments, direct messages, and live Q&A sessions, podcasters can easily connect with their listeners and build a strong community. This not only helps in retaining existing listeners but also attracts new ones through word of mouth.

Social media also offers valuable insights and analytics for podcasters to track the performance of their show. With features like audience demographics, engagement rates, and referral traffic, podcasters can gain a better understanding of their audience and tailor their content accordingly. This can lead to the creation of more targeted and relevant episodes, which in turn, can attract a larger and more engaged audience.

In addition to promotion and engagement, social media can also serve as a platform for collaborations and partnerships within the podcasting industry. Through networking with other podcasters or brands, podcasters can expand their reach and tap into new audiences. This can also lead to potential sponsorships and monetisation opportunities for podcasts.

Overall, the integration of social media with podcasting has opened up a world of possibilities for content creators and listeners alike. With its wide reach, interactive nature, valuable insights, and collaborative opportunities, social media has become an essential tool for the success of a podcast. As the industry continues to evolve, it's safe to say that the relationship between social media and podcasting will only continue to grow stronger. So if you're a podcaster, make sure to utilise the power of social media to take your show to new heights. So keep creating engaging content, connecting with your audience, and exploring collaborations on social media—the possibilities are endless! So make sure you don't miss out on this valuable opportunity for your podcast to thrive in the digital age.

Remember to always stay true to your unique voice and brand while leveraging social media for your podcast.

how much participation can you bank on for your podcast?

As a podcast creator, one of the biggest challenges you will face is how to get more people to listen and engage with your content. While there are many strategies and techniques that can be used for this purpose, one important factor to consider is the level of participation you can expect from your audience.

Participation refers to any type of engagement or interaction from your listeners, such as leaving comments, sharing your episodes, or even just listening to your podcast regularly. The more participation you have from your audience, the more likely it is that your podcast will grow and gain popularity.

So how much participation can you realistically expect for your podcast? Well, there is no definitive answer to this question as it largely depends on various factors such as the topic of your podcast, your target audience, and the quality of your content. However, there are certain steps you can take to increase participation and engagement from your listeners.

Firstly, ensure that you have a clear and well-defined niche for your podcast. This will help attract a specific group of listeners who are interested in the topic you cover, increasing the chances of them actively participating in discussions and sharing your content with others.

Another important factor is the quality of your content. No matter how niche or specific your podcast topic may be, if the content is not engaging and informative, it will be difficult to retain listeners and encourage them to participate. Make sure to constantly improve and refine your content to keep your audience interested and engaged.

Additionally, you can actively encourage participation from your listeners by including call-to-actions in each episode. This could be asking for comments, feedback, or even guest suggestions. You can also create a community for your podcast on social media platforms where listeners can interact with each other and with you.

Lastly, consistency is key when it comes to building participation for your podcast. By regularly releasing new episodes and consistently engaging with your audience, you can establish a loyal following who will actively participate and support your podcast.

In conclusion, participation is crucial for the success of your podcast. While there is no set formula for how much participation you can expect, by creating quality content, defining a clear niche, and consistently engaging with your audience, you can increase the level of participation and ultimately grow your podcast.

As Philippe Borremans, IBM's media lead in Europe, said in an interview, "Social media isn't for every company or every employee." But if you're one of the 60% of communication practitioners who now have some kind of social media program in place, what sort of participation can you expect from the workforce?

Looking at social media initiatives, including those featured in Melcrum's report, **How to use social media to engage employees**, the figures look like this:

- Around 4,000 employees of Sun Microsystems blog (11% of the workforce).
- At IBM, 23,000 employees are active on Facebook, and that was back in the mid 2000s; .

- At IBM's "World Jam"—an online multimedia brainstorming session in the early 2000s, 50,000 employees participated (14% of the workforce).
- At the World Bank during a recent leadership crisis, some 4,000 unique visitors visited the internal blogs (40% of the workforce), leaving 300 comments in 10 days.

Internal collaboration and knowledge sharing

ING group, a financial organisation with 113,000 employees, has an organisation-wide wiki—again to develop and enhance global internal collaboration and knowledge sharing.

The ING wiki was officially launched in October 2006. By March 2007 it had 2,200 registered users (2% of the workforce) and over 900 articles. The initial strategic goals of achieving at least 25 unique contributions per week and 400 unique visitors per day had also been met.

What does this mean for internal podcasting?

With a range of percentages from 2-40%, what constitutes a successful social media program? Of course, much depends on how long the program has existed. But considering the above together with more informal research and conversations, if 10-15% of your workforce is involved in any one initiative, it can be considered to be doing very well.

what about bandwidth concerns?

This is a standard concern for IT departments, and luckily the answer is a wonderful service called **Libsyn** (www.libsyn.com) where, for a small fee each month, you can host your podcast on their servers, set up the podcast ready to distribute via RSS and even retrieve download statistics on each edition.

A hosting package on Libsyn also comes complete with spaces and prompts for all of the essential information that you will want to make sure is present on your podcast's web page.

Podcasting offers all the key advantages associated with social media such as low barriers to entry and established popularity among the general population—in Melcrum's 2007 survey on social media and corporate communication, 58% of respondents said they were already using, or planning to use podcasts in their organisation.

HOW CAN I GET STARTED WITH PODCASTS?

Here are three quick stages to work through:

1. **Why podcast?**

First of all, consider what it is you want to use podcasting for. For example:

1. CEO podcasts
2. Daily/weekly team updates, divisional or organisational updates
3. Speeches or presentations
4. Training and administration purposes
5. Specialty podcasts
6. Something entirely different

Remember that podcasting is a technology that can greatly enhance collaboration. You could take this one step further and encourage participation and contribution by giving employees opportunities to participate in and develop their own ideas for podcasts.

2. **What to podcast?**

Next up, consider your content and format:

1. Will it be a program format?
2. Will it be news items, interviews, or magazine-style?
3. Will you have a single presenter or multiple presenters?
4. Will it be in a studio (if you have one) or will it be out and about? (Known as 'run and gun' in the industry)
5. How often will the podcast run and how long will it run for (a new regular show or just a series of six, for example)?

And finally...

1. **Keep it simple:** Don't be over-ambitious on your first few attempts.
2. **Keep it short:** Aim for around 10-15 minutes maximum.
3. **Don't overly fret about production values and technology:** A podcast that sounds like you're in a fishbowl will not work well, but equally, they don't necessarily need to sound like something produced by a professional broadcaster.
4. **Look out for best practice:** Always be on the lookout for new ideas and ways to improve the podcast—just as you would

do with anything else. That means taking suggestions on board and listening to as many different podcasts as you have time to.

5. **Keep them coming:** Don't use up all your energy on the first one and then give up!

PODCASTS IN ACTION

The next section of this report shares case study examples of podcasts and vidcasts in action.

case study: university of southampton, uk

ONE OF THE world's leading research faculties in digital media, the Electronics and Computer Science (ECS) school within the UK's **University of Southampton**, is one step ahead of the pack in its application of social media tools for communication. The school counts among its faculty the inventor of the World Wide Web, Sir Tim Berners-Lee, Professor David Payne, the inventor of the optical amplifier, and Professor Tony Hey, now corporate vice-president of Microsoft UK.

Since 2005, the school's lectures and seminars have been podcasted and latterly, vidcasted (video sent out using RSS technology, exactly like podcasts), allowing students to run them again for revision, or for the first time if they've missed them. A dedicated portion of the ECS website, www.zeplertv.com, was resourced to host the podcasts.

Experimenting with technology

In early 2006, the ECS communication team began experimenting with turning the technology to its external faculty news. "It seemed a great way to present our research to journalists in a visual and audio format," explains ECS's marketing and communications manager, Joyce Lewis. "It provided an easy and compelling format for them to digest, and a terrific way to provide a cognitive link for them between our research and its potential coverage as a TV or radio news item."

The benefits of podcasts

The result: "ECS TV" provides a vital link between the university's research work and the outside world. The benefits, says Lewis, are numerous—especially in a department of only one (Lewis herself), plus one web designer:

- **Low on cost, high on simplicity:** "When we first started it, we were able to do it just by using the video-conferencing technology we already had—fixed cameras, operated remotely, and radio mikes," she says. Editing the programs using the now-defunct *iMovie* application that came with her computer, the whole production line was cost-free. "That was the principal attraction of it in the first place—you can do this with a very small number of people and very few tools. It's high impact at no cost."
- **Speed:** The ability to turn around content quickly is an incredible advantage over previous forms of communication media, she says, and brings it closer to rivalling print. "We were able to cover a conference last year and podcast about it each morning for the entire four-day duration. Because the production is simple, it's really a case of a quick edit and you can get it out there."
- **Skillset:** With a designer who has a background in TV production and an IT staffer who had already (for the student lecture podcast series) developed the necessary compression tools to make the video RSS-enabled, "We knew we already had the skillset to do a pretty good job," says Lewis.

Discovering the speed and ease of podcasts

Joyce Lewis: Getting the best from podcast interviewees

Podcasts and video podcasts (vidcasts) are, very often, dependent on the quality and media-friendliness of their subjects. In a population of non-media trained people, however, this can prove difficult. Joyce Lewis, marketing and communications manager at the University of Southampton's Electronics and Computer Science school (ECS), has advice to share from producing dozens of such broadcasts over the last few years. The trick, she says, is patience and positive reinforcement.

If the interviewee is fine "off camera," she says, one must recognise that their stilted persona in front of the lens must only be a product of nerves and uncertainty, and can therefore be overcome with good, supportive coaching. Here are her tips for getting the best out of interviewees:

Take away the pressure

"If someone is drying up in front of other participants, I've found it effective just to ask everyone else to leave for a while—the camera person and other interviewees—so that I can sit and talk to the person and walk them through the process in private. That takes away the intensity and the pressure. Some people are shy and introverted and you need to help them relax. Taking away the sense of 'people are watching me' can help enormously."

Prepare them

In such a situation, Lewis will then often walk the participant through the questions and topics, and engage them in a conversation about the leading topics before filming begins. "If they've had the chance to articulate their thoughts on it out loud to me alone before we start filming, even just once, it helps them feel that at least they know roughly where their sentences are going when we do start filming.

. . .

Close the loop

A vital element of any communication is feedback, and this is doubly important for those who have been nervous about performing in front of camera. "You can do a lot with editing, so you can make them look as good as they possibly can. But I'll then also go through the video with them afterwards—because then they see that they're never as 'bad' as they think they were going to be, or think they were." So, it provides a vital opportunity to allay their fears, but also the chance to offer them encouragement. "Simply showing them the footage and saying, 'I think you were really good here,' 'I like the way you said that,' and so on will put them at ease. They'll feel much happier about the process and then they're much more likely to agree to do it again next time."

Joyce Lewis

Marketing and Communications Manager, University of Southampton, ECS jkl2@ecs.soton.ac.uk

case study: altana pharma inc.

Using podcasts to connect a remote sales team

HOW DO you connect with a geographically dispersed audience? iStudio, an interactive agency in Canada, introduced podcasts for one of its pharmaceutical clients as a personal and portable way to communicate with field employees.

Ron Clark assumed the role of Vice President, Sales of **ALTANA Pharma Inc**. The organisation needed a cost-efficient way to introduce the new VP, build rapport between management and their sales team, and provide insights on key business initiatives.

Introducing "Ron's blog"

The sales team at ALTANA is predominantly a remote workforce, spread across Canada. In pharmaceutical sales, representatives are primarily on the road for most of the work day. Accessing e-mail or the company intranet is often a challenge due to the constraints of travelling. iStudio investigated new online techniques that would be helpful in reaching a remote workforce and suggested a corporate blog/podcast as the solution. They created "Ron's Blog," which allows employees to have an open online dialogue with Ron.

Ron's monthly podcast is available via the blog. The podcast is a 10-15 minute audio file, which has a radio show feel. In each episode, Ron provides sales updates, shares insights and anecdotes and interviews special guests.

Early adoption success

The response to the podcasts suggests this medium has been a success: 76% of sales staff accessed the first episode and 88% accessed the second episode.

The feedback was also positive, with comments like: "It feels a little more personal than an e-mail, and has much more emotion attached to it than even a voicemail," and "I like it because it's innovative and user-friendly... plus I think it brings you closer to the sales force."

Clearly, employees find the podcasts much more personal than a voice-mail or e-mail. In addition, employees like the opportunity to provide input into the content of future podcasts via Ron's Blog.

Even though the podcasts are primarily intended for sales staff, over 80% of all employees access each episode.

To support the process of planning podcasts, iStudio has produced a customisable template that lets you plan podcast content down to the very minute (see https://bit.ly/3HspNvW). The template includes elements like "Call to Action"—a method designed to enliven the audio content and ask the listener to do something.

Fig 1: iStudio's customizable podcast template

Section	Time	Content
Standard Introductory Music	____ seconds	Theme music (check out the Podsafe Music Network: http://music.podshow.com/ if you need to select a song)
Introduction	____ seconds or minutes	Sample script or key messages (i.e. It's ____ (date) and this is the first episode of ______________ (podcast name). Describe focus of podcast and the topic for this episode.
Audio Transition	____ seconds	Theme music
Topic 1: (topic name)	____ minutes	Provide sample script and/or key points to be covered
Audio Transition	____ seconds	Theme music
Topic 2: (topic name) *Continue to add topics to fit the length of your podcast.*	____ minutes	Provide sample script and/or key points to be covered
Call to Action *(Highly recommended even if you are just asking listeners to provide their comments or ideas for future topics)*	____ minutes	Provide sample script and/or key points to be covered
Conclusion	____ minutes	Let listeners know when next episode will be posted
Total Time	**____ minutes**	

iStudio's excellent podcast template. Reproduced with permission.

case study: ibm

Driving efficiency with podcasts

"WE'VE SEEN a huge pick-up in interest in podcasting over the last 18 months," says Philippe Borremans, **IBM's** European new media lead. "Writing can seem such a technical skill, whereas people feel they can talk more freely than they can write." Borremans has overseen the development of a central podcast venue on the organization's website with all the tools, platforms and processes required to upload and share podcasts.

In the 12 months since its launch, it has become home to 304 podcasts, ranging from personal interests to business strategy.

One of the most consistently popular IBM podcasts, with over 20,000 downloads a week, is "IBM Week in Review"—a digest of the top internal business stories in IBM over the last seven days. Put together in a typical podcast style—a light, conversational interview between two members of the communication team—it's proving to be a particularly vital tool for remote workers, says Borremans.

"Because we have so many people working at client sites, they can't necessarily read all the daily updates on the intranet," he explains. "So, previously they would have missed out. Now, they can subscribe to a simple weekly podcast to give them all the latest news from the company."

One particular innovation—and lesson learned—from the process came from one of IBM's senior executives. His regular teleconferences with his global teams weren't working as well as he'd hoped—in part due to the time difference, with many of his direct reports having to dial in in the middle of the night; and in part due to the problematic feedback loop.

"On calls like that, it's hard to get true feedback," explains Borremans. "First, people have to respond on the spot, or the opportunity is lost. If something occurs to them after the call, it's too late. Second, it takes a certain kind of person to speak up live on the call, with everyone listening in."

The solution to these issues was for the vice president to launch the call as a podcast instead. The benefits are clear: his team can listen to it in their own time, rather than having to dial in at 3am. In addition, it means they can pause it if they're interrupted, or rewind it if they want to hear something again.

Driving efficiency with podcasts

Critical benefit: feedback loop

The most critical benefit however is in the feedback loop. The podcast is hosted as an entry on the executive's blog, which means that his team can post comments underneath the entry and get into a wide-ranging discussion about its content. And that's exactly what has happened.

"So even though a podcast is, on the face of it, more 'one-way' than a conference call, it has in fact revolutionized the amount of response he's getting," says Borremans. "He's finding it's really stimulating discussion about the content of the calls because his team now has time to listen to the call when their head is in the right place, and because they have a format—the written word—for giving considered feedback, rather than top-of-the-head, meandering thoughts."

Philippe Borremans: Key points to remember when podcasting

Based on his experience implementing podcasts at IBM, Philippe Borremans has learned a number of key lessons. Here he shares some pointers for communicators preparing to introduce podcasts into their organizations:

Understand the concept

A podcast is not just an audio file that you put online. "If you limit it to that, you're really not understanding it," he says. "The benefit of podcasting, as with a lot of social media, is that you're putting the user in control. If I subscribe to your podcast, I'm going to listen to your content when I find it, and when I have time—not when you upload it." That "time-shifted communication" is a core concept behind the technology for communication professionals to grasp.

Keep it regular

The most successful podcasts are those that form a series—so that people can subscribe to, and have a sense of continuity with, a new "episode" uploaded on regular diary dates. "It should never be a case of, 'Oh, instead of writing this, I'll just record it and put it online,'" says Borremans. "That doesn't work.

People are not compelled by one-off broadcasts. The whole point of podcasting is to create something regular that people will 'get into.'"

Subscription matters

Feeding from both the first two points, RSS is the vital glue—it's the subscription element that makes a podcast. As Borremans suggests: "So, you announce that every week or month we'll give you an update, which you can subscribe to by RSS—then you have a podcast. Not just because it makes it a series, but also because it will automate the download process—so that, thereafter, every new episode will automatically download to their PC or smartphone."

. . .

You don't need a computer

According to industry studies, 70% of people listen to podcasts on their smartphone, not on a computer.

Driving efficiency with podcasts

"It's a common misconception that your whole staff will need some sort of dedicated player for a podcast series to work. But you actually just need a PC or a smartphone," Borremans adds.

Learn the format

Listen to podcasts, understand how they work. They should be short (maximum 15 minutes), to-the-point and chatty. Also bear in mind why many of them are two-handers—it's much easier to listen to two people conversing than one person lecturing. "If you have a really good radio voice, you can do it alone," says Borremans, "but few people have that talent, so try to do it in an interview style. And don't script every single word or you'll have a very boring podcast."

Philippe Borremans

Europe New Media Lead, IBM philippe.borremans@gmail.com

case study: university of southampton, uk

Using videocasts to engage internally

THE ELECTRONICS and Computer Science (ECS) school at the **University of Southampton** oversaw the roll out of a podcasting initiative. Both podcasts and vidcasts were originally used to present research to journalists.

The video element appealed, largely because of the ease with which video clips can be generated. "When we first started, we were able to do it just by using the video-conferencing technology we already had," says Joyce Lewis, marketing and communications manager. Then the programs were edited using the *iMovie* application that came with her computer. The whole production line was cost-free.

Using video internally

Lewis realised early on that online video, in the form of video podcasts, had great potential internally. "Because of our core research areas, we have a good high-speed network and a lot of people interested in technology," she explains. "Plus, we realised that we could video things that our staff and students weren't able to see—like the development of our new Mountbatten Building complex that was, at the time, stuck behind a lot of construction fencing—so it had an immediate use as a visual communication tool."

"ECS TV" is a series of semi-regular video podcasts on a variety of topics. One "channel" of ECS TV is reserved for general ECS news (the updates on university research that go out externally to journalists, potential students, and alumni), and another for student news (for all students and potential students).

Powerful communication

But, of those run so far, Lewis feels none have showcased the medium's value as an internal communication tool quite so well as the third ECS TV series, covering the university's Mountbatten Building—a research facility that was destroyed by fire.

"The fire was a hugely traumatic event for everyone in the university," she explains. "Aside from the obvious, many students lost all of their research work in the blaze and the staff who worked there suffered three years' severe disruption of their research while the facility was being rebuilt."

One year on from the fire and, after protracted insurance wrangling, saw the final demolition of the remains. Lewis recognised an opportunity to use the event, and the anniversary date, to launch a special vidcast. Interviewing the head of school on-site as she oversaw the demolition, Lewis captured the raw emotion of what was happening. "By doing it with video—even a simple, handheld video with a small radio microphone—we created this enormously powerful piece of communication on the spot. She just talked very genuinely, unscripted, to the camera about what was going on and how she felt."

Video versus words on a page

It brought home to Lewis the power of the medium. "No matter how good the writer, words on a page from a communication department will never get close to conveying the sincerity and depth of emotion that one 10- minute video did," she says. "It wasn't staged, it was very intimate, so it felt like you were almost overhearing the head of school talking about it to you privately, and that made it all the more powerful."

Using vidcasts to engage internally

It also demonstrated the production benefits—the interview was completed at 2pm on the anniversary day and by 4pm the podcast had been uploaded.

One of the benefits of the medium's simplicity (and what allies it to the concepts of social media) is that it can showcase more direct and visible participation from the audience as interviewees. This is not lost on Lewis, who feels that it produces untold side-benefits for internal communication in general.

"If you ask people to be involved as interviewees, we've found they very quickly realise that what they say is going to be part of the school's communications," she explains. The result? You get both increased ambassadorship and a sense of ownership for the process. "So, they will say things like, 'Is this the right thing to say?' or 'Am I putting across the right message?' They clearly feel that they've become part of the communication process themselves, because they're representing the school and there's no intermediary between them and the messages.

"It's a very effective way to move towards a situation where everyone in the organisation is a part of the communication culture. And apart from the obvious benefits of creating a more two-way environment, I think that that gives people a feeling of real affinity with the work of the school."

On the production side, Lewis is also looking to roll out more independent video-making around the faculty, in line with this vision. "At the moment, we still control it. But from next year, anyone on campus will be able to make their own content to go up on the site as well. We'll still post it up, but the actual invention and work will come from them."

Encouraging contributions

To increase this participative, knowledge-sharing environment, her team has already been assisting people to contribute to the regular series. Students are encouraged to contribute segments to the ECS student channel since, as Lewis says, they know most about their own way of life.

One further instance on ECS News saw the showcasing of a team that had completed a research study into the future preservation of online media. Rather than produce the requisite written academic paper on the final findings of the work, the study team asked to do it as a video instead. "So, it was an ideal match of form and content, because of the subject matter," explains Lewis. Supported by her team, their work will form one episode in the upcoming schedule of video podcasts.

Video content can help move a text-only website from being a boring, static lump on the web to one that is more video-based, more interesting, and more engaging.

The idea is simple—icons at various points on the site, which, when the user clicks on them, open a small window with a video running of someone actually telling them about the site or some key piece of information, rather than the text-based equivalent. "Doing the video podcasts has shown us just how quick, cheap and easy online video can be now, and also its impact," she says. "With the way the web is going, things like that could soon be standard."

It's just one small concept, but part of an overarching theme Lewis sees as crucial to the development of web-based interaction for communication. "Rather than: 'What kind of things can we put on there?' I think we should be saying: 'We've got the web, it's an amazing multimedia source, in what different ways can use our sites to get our messages across?'

"If you think about it that way, then you think about the web more as one holistic thing, rather than just a place to put content."

Ripe for revolution

Lewis adds that the time is ripe for revolution: "The IT techie people have brought forward all the technology of the internet, but now we're taking it over as a communication medium. The technology is not that difficult anymore, the platforms (such as WordPress) are much more open to all of us. Now it's less about the technology and more about how we seize the opportunity to use it."

> **"No matter how good the writer, words on a page will never get close to conveying the sincerity and depth of emotion that one 10-minute video did."**

Joyce Lewis

University of Southampton

case study: ibm

Encouraging employees to videocast

A CORE PART of **IBM's** new-media strategy centres on video podcasting (i.e., vidcasting). "We certainly feel that there's a lot of interest in vidcasting inside the organisation," says Philippe Borremans, new media lead in Europe for IBM.

He suggests there are three core reasons the company has moved into video:

1. **It's how the web is going:** "Our moves to vidcasting really reflect what's happening on the web generally, where more and more of the content you're seeing these days is video content."
2. **The technology is already there:** With many departments, and most communicators, already having video recording capability, there is little extra cost to be outlayed. "And mobile technology is fast moving that way too," he adds. "I can take a video with my mobile phone that's good enough for a vidcast." The mobile technology means that there's also an increasingly strong, natural counter to the argument that vidcasts require devoted attention, which may mean less general productivity at work: "Just like podcasts," says Borremans, "people will be able to watch them in their own time."
3. **It offers greater creative potential:** With a podcast, there's a linearity to the content—15 minutes of material are 15

> minutes of material. But, with the right editing software, there are endless possibilities when it comes to creating communications that are inventive, clear and engaging. "Think about how quickly you can splice images, sounds, moving images, pictures and words in a piece of film to get across your message. You could condense the material from that 15-minute podcast into two minutes of video with the right suite of tools."

As it did with blogs, wikis and podcasts, IBM built an enterprise-wide vidcasting platform that centralised the technology requirements and keep content located in one place.

One of the core principles of the new-media team at IBM has always been to build social media strategy from the ground up—look at what employees are already doing in numbers (blogs, wikis, podcasts, vidcasts) and then respond to it with structure and the building of a central hub.

With vidcasting, the team wants to "keep the social media social." They therefore launched vidcasting with an employee competition—inviting staff to flex their creative muscles.

> "One of our core strategy topics is innovation. So, we invited all IBMers to create a short video clip to showcase how they see IBM innovating, or how they think it might help its clients innovate. They could then upload their clips to a central database, and have all IBMers vote on it for a prize."

The clips had to be no more than 60 seconds long and, in total, the company saw 512 submissions—ranging from the serious to the interpretive. (For example, one was a simple video about the potential of a paper clip). "It's one of those nice projects you can do to just have people think about a strategic topic for the company, but at the same time use new technology," says Borremans.

best social media platforms for businesses in 2024

IN 2024, the landscape of social media has evolved to include a variety of platforms that offer unique advantages for businesses. At the forefront, **LinkedIn** remains a vital tool for professional networking and B2B marketing. It provides a space for businesses to establish authority in their industry, share company news, and engage with potential clients and employees.

Instagram continues to dominate the B2C market with its visually-oriented platform, ideal for brands that thrive on image and video content. Its 'Shop' feature enables businesses to seamlessly integrate ecommerce into their social media strategy.

Facebook, with its vast user base, remains a key platform for businesses of all sizes. Its advanced targeting options for ads and the ability to create a dedicated business page make it a valuable tool for reaching a wide audience.

Emerging on the scene, **Clubhouse** has become pivotal for businesses looking to engage in community building and thought leadership. The audio-based platform offers unique opportunities for hosting discussions, panels, and Q&A sessions.

Finally, **TikTok** has proven itself more than a transient trend, particularly for targeting the younger demographic. Its short-form video content, coupled with creative and viral challenges, can lead to high engagement levels and brand visibility.

seo for newbies: kate toon and logan bryant

ALTHOUGH THE TOPIC of SEO (Search Engine Optimisation) is not really part of the remit of this report, what follows is a transcript of an episode of Kate Toon's superb podcast, 'The recipe for SEO success show'.

The reason I've chosen this episode of this particular podcast is simple: Kate and Logan are recognised experts in their field, they interact well with each other and each has many interesting and insightful things to say, they are an excellent example of how having more than one person in an episode lifts the episode to new heights.

In this episode, Kate and measurement maven Logan Bryant wax lyrical about the following items:

- What SEO testing is
- Where to start – simple tests for small business owners
- How to measure how successful your title tag is
- How long you should run a test before looking at the results
- How you can test the effectiveness of a featured snippet
- How to look at your highest conversion pages and see where the clicks are really going

- What tools Logan recommends for SEO testing
- Logan's top tip to get started with testing

You can listen to the podcast on the episode's page at https://therecipeforseosuccess.com/easy-seo-testing-tactics-how-to-get-better-results-with-logan-bryant-newbie/

Transcript

Kate Toon

Do you ever wonder how effective your site's SEO really is? You put all this time and effort into finding effective keywords, your images are slim, and your page is loading faster than a fast thing. But is it bringing in the clicks and the sales? Because at the end of the day, while ranking is nice, we know that what it's all about is making money. So today, we're going to look at some tests you can run on your website for SEO to make sure it's all ticking along beautifully, and what needs some work. Hello, my name is Katie. I'm the head chef at the recipe for SEO success and online teaching hub for all things related to search engine optimization, and Digital Marketing. And today, I'm talking to Logan Bryant. Hello, Logan.

Logan Bryant

Hi Kate, how are you?

Kate Toon

I'm very well, it's lovely to have you here. Let me tell everyone who you are. So Logan has been building content driven SEO programs at companies like to polti workboard Koala and clearlink. Early on in his career, he became obsessed with the question of which activities work versus which activity stones. This turns into a natural curiosity that has led him to develop an experimental mindset. In most cases if a thing is worth doing, it's also worth measuring and learning from Logan likes to combine curiosity with measurements to untangle the messy black box of SEO efficacy. Fun fact, I love the fun fact despite being a notoriously picky eater, Logan has eaten ballute duck embryo and he likes it. Okay. That's a bit weird. And I'm a vegetarian. So I think maybe we should just I'm not sure what. I'm not being funny. But how does one come to eat a duck embryo? It doesn't seem like a regular thing that will be happening in the middle of America.

Logan Bryant

No, definitely not. I spent a couple of years living abroad in the Philippines and they had that there that's like where they're notorious for having the balut. And I it took me a year I was there for a whole year before I even went near it. And it honestly it wasn't as bad if at least you know, maybe if you're a vegetarian, don't don't go Yeah,

Kate Toon

I don't think it'd be top of the list. Did it? Was it the classic thing of everyone says when they eat something weird? Oh, it tastes like chicken? Did it taste like chicken?

Logan Bryant

It absolutely tasted a bit like chicken. Because it's so, duck isn't really that far. Yeah.

Kate Toon

It's like a chicken cousin, isn't it? Yeah. I mean, I remember going through the markets in Thailand. And there's all like, sort of, you know, battered tarantulas and locusts and all kinds of things. I was very grateful at that moment. I was a vegetarian, and I didn't have to be adventurous. So there you go. But anyway, when it comes to SEO, we should be adventurous. And we should try new things often. Look at that segway Logan. And that was pretty good, right?

Logan Bryant

Not bad. Hard to go from duck embryo to SEO.

Kate Toon

Duck embryo to SEO, there's always a connection, I can always find it. I think when people, you know, they do their SEO things, and they kind of they'll open up their Google Analytics, I'll open up the Google Search Console, have a bit of a poke around and go look, okay, then, and make absolutely no changes at all. So when we're talking to complete newbies, what do you really mean, when you talk about SEO testing? What sort of things are we testing?

Logan Bryant

Yeah, so one of the things I think people's brains sometimes I've noticed tend to shut down when you when they hear SEO testing. Oh, it's this big advanced thing. And they start thinking CRO, and they start thinking about hot jar and, you know, Optimizely, and all kinds of different testing stuff, and it starts to just be overwhelming. But really, it shouldn't be that overwhelming.

The best way that I like to think about testing is just break it down, just scale it back to what the word testing even means. You know, at the end of the day, it means trying something. That's what a test is, whenever we're testing something, we're trying something. And at its most basic, fundamental roots. That's what I think testing is. And you don't have to have a perfect process, you don't have to have the most scientific or statistically significant method to your testing, aka trying approach. You just have to be willing to try things and at least have some idea of like a hypothesis or an idea that you want to try and then also just some way of measuring it even if that way isn't perfect.

Kate Toon

I love that I love the being able to release the imperfection and also focusing on small things, you know, one thing at a time, and I like your idea of creating a hypothesis because I think people often go into testing tools that are available with no idea in mind of what they're looking for. But if you're like right, well, I'm gonna try and I'm sure you're gonna give us some examples. I'm gonna try So I've been writing 1000 word blog posts. And from here on out, I'm gonna try pushing up to 1500 words and see if that improves my time on page, see if that improves people going from page to page. And I'm going to try that for a period of time. But let's, let's break that down. What if you were starting? If the first test that a small business owner could do on their SEO site? What would you recommend? What are some simple tests, you can run for SEO?

Logan Bryant

I tell pretty much everybody to start with titles, that tends to be the first test for most people that are attempting to impact their SEO, if you think about it a title tag, it's not very long, they're usually somewhere between maybe 30 to 60 characters somewhere in that range most of the time. So it's really not that long. And it's not very high effort, as you know, in comparison to writing a whole new blog post, yeah, all you've got to do is just have an idea for, hey, this title looks like it could be improved in a certain way. I typically like to analyse what's on Google to see what other people are writing and then see if Hey, maybe there's a way that I can improve my title, then you just launch that title test and see what happens.

Kate Toon

Yeah, and I like that. I think people are quite terrified about touching their titles. But the truth is, Google manipulates them anyway. And often, one of the things you can do with title tags is look at what how Google has changed it based on searcher intent, look at the ads that other people are using, if there's ads that come up for your search to see what phrases they're using, and then give it a change. And then what metrics would you use to measure whether you've made a good decision or a bad decision with your title tag?

Logan Bryant

Yeah, and if you don't mind, I'm gonna go on a very short tangent on that. You. It'll be short, I promise. But if you if you take the title tag with you mentioned dynamic title rewrites from Google 9.9 times out of 10. When Google changes it, it's usually because it likes your h1 better. So a fun fact, if you're doing a title test, and you're thinking about the dynamic stuff happening, if you change your h1 to match the new variation that you want, you have a better chance of Google selecting that new variation. Because when it goes to the to the h1, it'll see the same new variation that you're wanting to test.

Kate Toon

That's not a tangent. That's a bloody excellent tip. I love it. Fantastic. So title tags, obviously, one of the you know, the best things that title tags do is encourage click through rate. So you could look at your click through rate before, and then look at your click through rate after and see if that's improved, because you know, obviously, it's improving people's ability to understand your content. Could you know what else would you be looking at to see if you've done a good job or a bad job?

Logan Bryant

Most of the time, with a simple test like this, we're looking at before and after data. And I like to look at just overall clicks, you know, you can look at click through rate data as well. And it is helpful to look at multiple metrics in, in, in union with each other. But I do like to look at clicks as the overall benchmark as to whether or not I made an impact or didn't make an impact, because that's what we're looking for is clicks. So most of the time I go for clicks. However, there is another scenario in which rankings become very important, as well, because sometimes you might have a page or a URL, and that URL may be ranking on page two or three for a keyword that you're really trying to rank for. In that case, measuring clicks isn't going to be super helpful, because you're not going to be getting a lot of clicks to that URL in the first place. So it is definitely not uncommon to run a title test on some on a page like that. See rankings go up and be able to measure an impact without any clicks data whatsoever. And that's another reason why I tell people, they don't always have to be statistically significant because rankings when they go up, there's no statistical significance there that you can really measure. But is it a valid test? I would say? Absolutely. It is, especially if you see rankings go up.

Kate Toon

Fantastic. All right. Well, that's one really simple test that we can do, which I hope everyone will maybe go out. And that, you know, obviously what Logan saying is No, don't go and change your titles on 200 pages, you know, pick two or three products, maybe one that's or pages, one that are doing well already, could you improve it a little bit and one that's languishing, could you improve it a lot? I think there's a simple test. And I guess the other question would be how long do we wait logon before we kind of look for results, because people are very impatient. And you know, we know that with Google, you know, doesn't crawl everything instantly. And it may take a little while to see results. So for tests like that, how long would you wait before you started looking at results?

Logan Bryant

Yeah, I typically advise people and you'll hear a lot of other people Both who are running SEO tests stick within the two to three week timeframe. And it's important to note that that two to three week timeframe, it's not really a magical number that that tells us Hey, and in two weeks, that's when we know. So it's not, there's no set magical timeframe that you should always abide by the however, the reason I, the real reason I go with two to three weeks is mostly because I like to be running multiple tests. And I like to run subsequent tests after my initial ones, which means that if I want to run more tests and kind of create a backlog of testing ideas and get through those, I like to move through those at a cadence. And that two to three week timeframe makes a lot of sense for me to be able to continue testing in a in a more structured way. So you can see results earlier than that, you might need want to wait longer than that. But it's a good idea to give it at least two to three weeks.

Kate Toon

And I think that's often the problem that I see with the students on the course is once they get the SEO knowledge, I want to do everything at once. And I'm often like, you know, be iterative with it. Because you will not know if you're you know, you've fixed this and you've changed that you've changed that. And then you see a change in results, you will not know what was the deciding factor? Was it that you did x? Or was it that you did y and you've got too many tests going at the same time. And they're all impacting obviously, everything's connected, you know. So having that kind of iterative kind of cadence, waterfall approach of one thing than another. And having a little bit of patience can be better. Because obviously, if you find that, you know, you have made a correction to a title tag, and then that's something that you can then write and it's made an improvement. And that's something you can roll out over other pages. That's great. That's great. You've learned from that, and then you can roll it out. But doing too many tests at once can be a bit overwhelming, don't you think?

Logan Bryant

I think you can do a lot of tests at once. As long as they're not overlapping. That's it. And it can be overwhelming indeed. So you know, I wouldn't say overdo it, but go at a pace that's comfortable for you. And that aligns with your goals that you have, whether they be growth goals or educational goals.

Kate Toon

Right? Well, before we move on to talking about like, you know, using data to improve traffic and sales, can you maybe give us an example of one more test? Because I love that one? It's so straightforward. One more test that someone could try this week on their small business website.

Logan Bryant

Yeah, my next favorite most favorite test is featured snippets. Oh, so usually people think of these things, such as featured snippets, or even content refreshes. A lot of these projects that I think tend to be labeled as optimizations. I don't always think of them as optimizations because at the end of the day, how was it an optimization if the result didn't follow? So experiment? Yeah. Yes, exactly. So a Featured Snippet test. The really interesting thing about those is the result is binary. So you either get got the featured snippet and one that result or you didn't.

Kate Toon

And why anything in SEO that's actually black and white is such a relief, isn't it? Because everything's such a sliding scale? So it's either a love a yes or no answer. So what would you how would you go about testing that? What would you do? So say, you've got an existing blog post, for example, and you're hoping to get the featured snippet for a particular question? What actions would you take there?

Logan Bryant

Yeah, most of the time, you're, first of all, you want to make sure that you're ranking somewhere on page one, if you're not on page one, you don't, you're not, you've got to get it ranking higher to even be in the running for that featured snippet. That's the first criteria. And ideally, you want to be somewhere in the top five, you can absolutely still win a Featured Snippet if you're in position nine, or 10. But your chances decreased quite a bit. So hopefully, you're on page one, and hopefully, you're even higher up on page one already to begin with. Once you've done that, I like to just check out first, like, learn from what Google's selecting because a Featured Snippet test, you're kind of trying to predict what Google thinks is the best answer. So in a lot of these answers, there's not an objectively best answer. It's what Google's algorithms think is the best answer. So learn from whatever's in the current featured snippet. Make sure you're matching the format. If the format is a ordered list or an unordered list or a table, make sure you're matching that format. Most of them are going to be paragraphs. So nine times out of 10. That's probably what you'll see as a paragraph and then try to rewrite that same answer in a way that doesn't exactly match the answer that is existing but you might want to just do a spin off of what's existing. And see if you can improve on whatever answer there is. Make sure you're staying succinct. Make sure you're not using any filler words or fluff and answering in the most direct way as possible. And of course, we could go a lot on a lot more Featured Snippets. But that's, that's essentially how I would do it. And then I would just treat it just like a title test, where I track the what I had before. And I track my new answer. And I see which one Google preferred.

Kate Toon

I love that. I love that. And, you know, a simple way that I do that is in my my blog posts, I kind of answer the question shortly. And then I answer the question Longley, as I like to say, so a nice little snippet at the top that summarises the answer. And then you can continue on in your piece of content to expand on that and add more detail and videos and all sorts of things. But yeah, having a sort of shorter answer at the top of your content can be great for Google to kind of scrape and use, but I love those tips. Fantastic. So if we are sitting here and listening to this and thinking these both sound great, you know, obviously they're going to drive more traffic to our sites. And he's going to be more qualified to drive traffic because people are interested, the title tag is more relevant. The Featured Snippets partially answered the question shown or authority, and we were keen to click through, what other kinds of data can we use from testing to increase our site traffic and sales?

Logan Bryant

Well, I'm not sure in terms of data, because I think that's most of the data other than if you want to look at conversion data, which you know that conversion data tends to happen, after we generate the traffic, which SEO testing doesn't go as far as to dive into the actual conversion rate optimisation tactics. So usually, the goal of SEO testing kind of the furthest it can take you is traffic growth. So I don't know that there's a lot of other data. But I will say that the best way to utilise it is to focus on your highest converting pages focus on your money pages. So you can look at existing conversion rates to see which pages are currently converting. You can look at striking distance, come to think of it, there are a few data points. One of my favourite ones is I call it 'Available Clicks', which is where I look at all the URLs on my site. And I will, I'll pull the Search Console data. And I'll say, what I'll subtract the clicks data for each URL from the impressions data. And that tells me how many people are getting an impression but not clicking through. And I'll sort that column to see where are the most available clicks, leads me to some of the best SEO testing opportunities because it tells me, hey, if my title tag performs better, or my Featured Snippet performs better, I'll generate even more clicks off of these pages, and I can prioritise my efforts in that direction.

Kate Toon

Yeah, that's such a great idea. I've not heard that one before. That's amazing. And I think you're right, you know, I like to say SEO gets people to your door, and then your content and your CRO drags them through and gets the sale. But I guess you could also look at, you know, new, new versus existing traffic with the basic premise that people who've been just before a slightly more likely to buy than people who've never been to your site before. So what could you do with your testing to encourage loyal customers to come back again? And I think one would generally say that, even when you got the traffic engagement, is it people are more likely to buy if they're engaged with your content, if they're spending more time on the site, if they're looking at more than one page, if they're getting further down the page? So you possibly could look at some engagement elements as well, don't you think?

Logan Bryant

Yeah, I think you can, once you get into engagement elements, then I would suggest doing tests that are a little bit more related to the site content after they click yes, because the title tag is usually kind of like, after they click, they don't see the title tag anymore. The Featured Snippet. Yeah, they see that, but it's only a small part of their overall engagement. So I would think about doing like a content refresh test, like hey, can we redesign it? Can we refresh it? In those instances, when we're trying to improve those like, in on site engagement metrics, you might get a little bit more thorough or a little bit, you know, more hands on with larger tests.

Kate Toon

Yeah, I mean, even just basic things, you know, working with some of the E commerce stores or just taking the best performing products, you know, that are already getting good clicks and good traffic's and just augmenting them adding a bit more detail, maybe adding a video, adding some FAQs, adding a couple of additional pictures just to get people to get more engaged with that content on the page, which doesn't sound like SEO, but it can Have is SEO. And it's because this is I think people try and be too distinct about oh, this is SEO. And this is CRO, and it's kind of all combined to some degree, I think. So we've talked about some great tests, as we talked about Title Tag testing Featured Snippets. Looking at conversion data, I love the available clicks tip that's just fantastic. Looking at the gap between impressions and clicks. That's amazing. And then some engagement factors. But what tools would you recommend for people to use to do these kinds of tests? So we are we sticking with Google guts Search Console, or using one of the sexy expensive ones? What what do you use to do tests?

Logan Bryant

Yeah, I was think if if you're just starting out, just use Google Search Console, if it's your first test, if you're still getting in, get, you're getting your hands dirty with SEO testing, you don't need to sign up for any tools, I would just use Google Search Console, maybe some of Google Analytics, although I prefer Google Search Console in these specific testing environments. If you do want to start doing more testing, meaning, hey, I want to scale up to do more tests, or I want to measure my tests better. Or I want to do even more advanced testing. Like there's there's SEO, split testing, which we haven't talked about, but we shouldn't, because it's very advanced. If you want to start going in that direction of higher velocity or complexity of your testing, then that's when you would want to look at other tools. And the first one I would recommend graduating to would be SEOtesting.com. That's a great tool. It's not very expensive, I think, I don't know the exact pricing. And I'm, by the way, I'm not sponsored by them or anything like that, but but I think it starts at like $30 a month or something like that. So it's really not too bad. And they can give you that tool can give you a lot of data, and it can really help support and help you scale out your SEO testing program. So that's where I would start, there's also really advanced ones for enterprise like search pilot is a fantastic, really cool tool. Once you get to some more really advanced testing techniques, there's several others out there. But I've also got a you know, maybe a selfish plug, I've got a tool that I've been building, it's not launched yet. That's basically just designed to help scale out more testing as well. But

Kate Toon

if you've got a wait list going for that, we'll drop a link into the show notes.

Logan Bryant

I do actually it's testlab.com.

Kate Toon

test lab.com, we'll put that in the show notes. So go to the 'Recipe for SEO Success' to see that. But I love you know, I do think that for most, you know newbies or beginners or even people who've been around a while Google Search Console is great, we've got all the data there, if you go into your, you know, performance on search results report, you can see clicks, you can see impressions, you can see click through rate, you can see average position. And that's a great place to start, especially with things like those title tags. And then obviously, using the search results themselves, I mean, you know, obviously, they're going to be to a degree personalised, but you know, the Featured Snippet one, it's either there, or it isn't, you know, it's either there or it isn't. So you can actually just look at the data within Google itself. You don't need to go out and necessarily pay for a sexy fancy tool. You know, on top of that, unless it's meta labs, meta test labs.com, then get that one. So look, any final tips for people who are thinking about starting out with testing, you've given already given us some great examples, any other tips that you can give us the wrap up the episode?

Logan Bryant

The main tip I would want to leave people with is learning how to think with an experimenters mindset. So more than anything, don't overcomplicate your SEO testing efforts, and start thinking like an experimenter going back to the idea that SEO testing is really just about trying new things, and giving things a shot and seeing what happens and measuring those results. Stop thinking about optimisation so much, because you don't really know if it's an optimisation if it doesn't work. So think a little bit more like an experimenter. And I think that will open up your approach to SEO in general, not just SEO testing.

Kate Toon

I just want to finish off on a final point, because when I was, you know, learning SEO 10 or so years ago, I use my own site as a petri dish. So I did all my experiments there, you know, SEO was kind of in its infancy back then. And we had some information from Google, but not much. Those were the days of Rand and his whiteboard a long time ago. And I was unafraid of, you know, fiddling with my site, you know, I was happy that it would have wins and I would have losses, but a lot of people will be thinking like I don't like I don't want to change the title tag for this page. Because what if I flick it up? So you know, can how can we get people away from the fear of this and that they're not going to break their site by change? Take a few title tags.

Logan Bryant

There are very, very few instances where your SEO test is irreversible. Yeah. Almost everything is reversible. If your title tag doesn't do as well, if your featured snippet doesn't work, if your copy does poorly, almost always you can revert right back to what it was before. And Google will pick that performance right back up, and it won't be a big deal, I promise. So that's the number one thing. Just don't be afraid, because you can always go back.

Kate Toon

Yeah, I love that. And you know, we're not talking about doing things at mass scale, changing 1000s of title tags on a single day building 1000s of links in a single week. You know, it's small, iterative changes that then you roll out over time. And as you say, Everything's undoable. That's the great thing about the internet. It's not like print. Back in the day when you make a mistake there. It's there forever. So everything's fixable. laughable. So Logan, you mentioned your new testing tool, meta test. lab.com. Where else can we find out about you? Well, which social media channel do you hang out on the most?

Logan Bryant

I am most often on LinkedIn, I occasionally pop into Twitter/X now.

Kate Toon

X, yeah, wild west land there. Now, I found Logan via LinkedIn. And that's why I invited him on the show post lots of useful tips and advice. I will include a link to your LinkedIn on the site so people can go and follow you and read all your bits and bobs. And we'll also include links to Logan's website and his Twitter strike X account, as well. So Logan, thank you so much for coming on and sharing some really practical doable tips. I really appreciate it.

Logan Bryant

Amazing. I hope it helps. Thanks so much, Kate, for having me.

Kate Toon

That's the end of this week's show. If you have questions about SEO testing, then head to my 'I love SEO' group on Facebook, I'm going to do a little demo of a couple of Logan's tests because I thought that was such great advice. Also, if you don't have it already, check out my free ultimate SEO checklist. That's a great way to start with testing. Go through that and make sure you can tick every box. And that's a great way to get started feeling confident that you can make changes to your site and that you're not going to necessarily break it. Anyway, we'd like to end the show with a shout out to one of my lovely listeners. And today it's Leesy-lou from Australia. Brilliant. There isn't too much that this lady doesn't know about SEO, love her knowledge, content and podcast highly recommended listening 10 out of 10. Thank you, Leslie Liu. And thanks to you for listening. If you have a minute to leave review wherever you're listening to this right now, it's a bit harder on Spotify. But on iTunes. It's as simple as clicking on the app and clicking on the five stars. You don't even have to write anything if you don't want to. If you do, you'll get a shout out on the show. And don't forget to check out the show notes for this episode. I'm going to drop in a link to Logan's tool and to the other tools that he recommended. You can also leave a comment about the show. So until next time, happy SEOing.

Connect with Kate

therecipeforseosuccess.com

clevercopywritingschool.com/category/podcast

Connect with Logan Bryant

• Logan Bryant [loganbryant.com/]

• Metatest Lab [metatestlab.com/]

• LinkedIn [linkedin.com/in/loganbryant/]

• Twitter/ X [twitter.com/LogansNotions]

Useful Resources

- Google Search Console [search.google.com/search-console/welcome?utm_source=about-page]
- SEOtesting [seotesting.com/]
- SEO pilot [seopilot.io/]

steve davis answers questions about podcasting

HAILING FROM ADELAIDE, South Australia, Steve 'Mr Podcast' Davis is a consummate podcasting professional—like me, he has been podcasting since the beginning of social media podcasting (2005), and for ten-plus years has published the very successful and award-winning podcast, 'The Adelaide Show'.

By day, Steve is a successful marketing resource for small and medium-sized businesses, offering those fortunate enough to be his clients practical and strategic perspectives, insights and how-tos for taking their business further and higher, whilst ethically generating more revenue and greater profit. 'Talked About Marketing', the podcast he publishes with his colleague David Olney, encourages listeners to put aside the day-to-day and instead focus on the bigger, more high-thinking picture. It is compelling and challenging listening, and a 'must' for business owners and fellow marketers looking to make a difference in their clients' lives.

Steve recently recorded a video on podcasting, answering a student's questions about what makes for a good podcast.

The astute amongst you will note that Steve has a very polished and professional voice, no doubt honed by his twenty years in the radio broadcasting game.

I mention this because I need to make a very important point:

> **You don't need to have a polished voice to be compelling. It is the message you have and the story you tell that makes you a good podcaster.**

Steve is blessed with having a 'good' voice, but that is not why his podcasts are successful; if they are successful in any way it is because of the content he provides in co-operation with his guests, and the stories that he and his guests tell. Don't be put off from starting a podcast if you think your voice is not 'tv or radio quality'. Focus on your story and your content, **not** your voice. Over time and episodes of your own podcast, your voice will become more polished, trust us.

Transcript

It is the 10th anniversary of the Adelaide Show podcast, we are halfway through our month of Tuesdays, and I've just received an email from a student in a South Australian school. I won't mention names because of privacy, etc., but they're doing a podcasting project, and they've asked me a couple of questions... well, 13 questions to be precise.

1. How do you create a podcast, do you just record or is there a need for editing?

That all depends. At the base level, yes, you just need to record. Obviously, some more thinking needs to go into it than that, and that's going to be covered in a later question. Editing is a good thing, because if you're new to this and you're waffling a bit, or you lose track or you're if you have a guest if they go off track you want to cut that stuff out and keep the essence of what they say. Probably the biggest thing that I would encourage anyone to do, is if you have a guest on for your podcast don't just say "right, here's our guest, tell us your background..." No, that's *your* responsibility. You give us a succinct version of the background and if you want to maybe ask a specific question that relates their background to the topic of this discussion, and that just makes it better for the listener.

2. How fast should I be talking?

Well, probably about the pace I'm talking now is going to suit most people. I wouldn't go much faster than this. You could go slower, but I think if you can keep this pace in your conversation there's enough energy to keep people connected, knowing that many podcast listeners play podcasts at a slightly faster speed. I personally don't like that, but some of them do.

3. How do I keep listeners interested?

Well, editing is going to help you because you cut to the chase. Try to open up with an intriguing hook as to what's coming up and then work your way through that material, that's a good ploy, and stay on track. In fact, it really helps if you can imagine what an ideal listener is like; just pick someone you know that you think represents the person and picture them or put a picture of them in front of you as you do your podcast. This will keep you focused, remembering they listen to this in their own time, they have to want to find it and listen, so we owe them. That'll be a good start.

[Question number four was left out due to perhaps typos or just a software glitch—Lee]

5. Do you have any expert tips?

Well, the intro thing I just talked about is one expert tip and... look, map out some questions and think of a logical progression that you can go through with your topic, but don't just do the questions on their own, it takes a bit of practice. I've done more than 50,000 interviews in my career; having that background is nice— it's like having a general idea of how to get across town in your car—but every now and then there's a detour or there might be an interesting deviation you can take; go for it.

So, two things: Listen as the person's answering the question; if you're doing an interview-based podcast and if there's something you don't quite understand, pause and ask them, "So, what do you mean by that? or "How do you define that?" because one of my bugbears is listening to people doing interviews where they ask a question and then once the person being interviewed stops talking they just asked the next question on their list, and there might be some really interesting stuff left on the table that is just ignored at that point.

Make sure the room you're recording in is quiet. Make sure they're close to the microphone, because the further away the more you're going to get that boomy, echoey, noisy sound.

6. Ideal podcast length?

Somewhere around the 20-minute mark is pretty good for an episode. Some do shorter, some do a lot longer; 'The Adelaide Show' is a lot longer, whereas with the 'Talking About Marketing' podcasts we aim for 30-ish minutes. There's 'This Medical Life'—30 to 45 minutes all up. I do a comedy one, 'Hard Knock Knocks', that's about 30 minutes. I think you find 20-ish minutes is about the average around the world.

7. How can I make it interesting?

It all depends on your topic. I think you've mentioned there's a fish aspect to what you're planning to do a podcast on it in your case. So, why would someone care? What's in it for them? How are you going to surprise them with something they'd like to know? What you're doing —whether you're just sharing stuff you've learned or whether you're interviewing someone—you're there on their behalf to try and get some interesting things to help them either have fun or in this case, probably more to the point, learn something about the industry. We don't get to sit down with people who know their stuff very often, so you're providing that service, you're giving us a chance to have an expert sit with us and speak into our ears, give us an insight. Stay on that focus.

8. What should the podcast structure be?

The intro, the info, the ending but that is a big question. It all depends. I do like having a structure, so for 'The Adelaide Show' we have the 'SA Drink of the Week' as our first segment, always. I do a very quick intro to tease what's coming up, then the wine tasting segment, then our main interview, and then we always finish with a song by an original South Australian artist or by an artist an original South Australian song, you know what I mean. So, that's my structure for that one, for others we come up with something different.

For 'Talking About Marketing', we've got four segments we do every single episode so the topics change but people get to learn what the structure is about. So, intro—just tell us up front what's coming, don't have a long build up to it, cut to the chase: "Hey, in this episode you're going to learn about the three biggest threats to South Australia's fish breeding industry, or whatever it is that you're talking about, so bang you've got our attention, now we're going to listen to the info.

I think it's covered in the guts of what you do, and the ending let people know where they can find more information or pose a question for them to ponder and maybe respond to you to for their comments to be included in an upcoming episode, if that's how it's going to work; just know there's not a lot of that happening in Podcastland because people are often listening when they don't have a ready access to make a note or remind themselves and then they forget, they may want to respond to you but it's just not the nature of the beast, typically.

9. What are the steps to creating a podcast that will work out?

The show, the structure. First, get some recording gear: I just have a zoom H5 recorder, but any good recorder will do. You can record straight into a computer as well. The microphones are important—I use the Shure SM58, they're my microphones of choice, but I've got some good lapel mics that I use that I plug into my smartphone, they record sound well. The key thing is having some good sound. And then there's editing: I just use **Audacity** which is free, open source software. [So do I. We've both been using it forever – Lee]. You can put it on a Mac, you can put it on a PC, it does the job. It cuts out the stuff you don't want, lets you shift stuff around, and so on and so forth. They're probably the two things I'd recommend. Then, of course, I export to MP3 and the highest bit rate possible—I think it's 320 kpbs—doing that just gives a nice round sound to the MP3 file.

And, of course, the other part of it is you need to upload it somewhere. If it's going to be a true podcast, you need a host to hold those files then serve them out and make sure they're available through Apple podcasts, through Google podcasts, through Spotify, etc. A good one to start with is **Podbean**, they've got a free entry-level plan to get you started and then about US$8-9 a month for more advanced hosting, which is not a bad way to go. There'll be some other ones poking around, they're changing all the time, but **Podbean** have been around for ages.

Once that is there, **Podbean** will teach you how to connect your podcast to all those broadcast avenues, and also don't forget to upload it to YouTube as well. I haven't been doing that recently, I'm about to start doing it more as there's a lot of people out there who only like listening to podcasts on YouTube, so either video what you're doing and have it up there or just get your sound file, throw it into something so it can have an image (just a still image and the sound) and get it onto YouTube because you'll pick up some extra listeners that way.

10. When and where are sound effects useful?

Depends on the style. I don't know that you're going to need them if you're doing one of those podcasts of just people sitting around a table chewing the fat, but maybe some podcasts.

We played with that in the early days of 'The Adelaide Show' just for a bit of a laugh, but the novelty soon wore off. That said, if you're out and about in nature and doing some recordings, softly having some atmospheric background noise of—in this case with fish bubbling tanks or whatever—is good. Too much, though, becomes annoying, so experiment see how you can add some character and some richness to the sound, but doesn't dominate.

11. Would the video editor on Microsoft Windows work?

It can be useful for creating a podcast, but **Audacity**, which is the one I mentioned before, is the one I recommend.

12. Is humour useful for podcasts?

Again, it depends, I think. There's always a place for humour; going out for the sake of making a joke, though, is a hard gig. There's a natural chemistry that makes most comedy work, which you don't always get in a podcast environment unless you're sparring and bouncing off a partner or a guest, so it depends on your character. There's no right or wrong answer there, but just think about it—most conversations with humans always have a little bit of a joke along the way and that just adds a bit of smile. A bit of sparkle keeps us engaged.

13. What else should I include in podcasts other than the information?

Your personality, what's your take on it, how do you connect the dots between your lived experience and the topic that you're covering? Why do you care? Sometimes grounding your experience so you're surprised by what you're learning, or you can add an extra bit of spin, an angle, to a conversation, to a question that's useful—you're there partly with people who listen to podcasts, they build an affinity, they build a relationship with the podcaster and it's a very intimate thing, we're in someone's ears and it's just our brains which have evolved slowly, that's still creating one side of a relationship, so it's a lovely place to be, it's an honour, it's a privilege.

On that front, what you should include is a little bit of yourself, and part of that is don't be afraid to ask the dumb questions if someone says something you don't quite understand what they mean, ask them and be a little bit dogged, don't just let them get off the hook (hook, fish... see what I did there?) See it through to the end, not to be nasty about it but just if you're driven by curiosity. If you can bottle your curiosity and honestly want to learn, I think that will make a good podcast for everyone.

The podcasts Steve produces

theadelaideshow.com.au

talkedaboutmarketing.com/podcast

thismedicallife.com.au

ncver.edu.au/news-and-events/podcasts/vocational-voices-podcast

hardknockknocks.com/podcasts

Products that Steve mentions

Audacity, the world's most popular audio editing and recording app [audacityteam.us]

Podbean podcast hosting – Discover all the features you can use to publish, promote and monetize your podcast [podbean.com/podcast-hosting-features]

Steve's video on podcasting from which the transcript was taken [youtube.com/watch?v=UbjZXgC697w]

how to achieve audio excellence in podcasting — neville hobson

I FIRST MET Neville back in January 2005, when I was pointed to the podcast he and colleague Shel Holtz ran (and still run!). They were the first podcasters in our Business Communication industry, an industry which included PR practitioners, marketers, public speakers, trainers, and the almost unheard-of cadre of corporate communicators.

After submitting a few spoken sound files to them, complete with sound effects that possibly conjured up mental images of Australia and its fauna, Neville and Shel kindly appointed me their first Foreign Correspondent. A happy union was achieved, and I kept submitting 'reports' until my business life got in the way and my social media consulting work meant that client work had to come before playtime with **Audacity** and my role as a correspondent.

All three have kept a watchful, interested eye over each other's work—blogging, micro-blogging, podcasting and vidcasting.

Neville always has something interesting to say, some new bit of social media technology to have a look at, some new insights into what has and continues to shape the social media world, both for businesses and consumers. He is especially gifted in peering into his crystal ball and determining what is ahead and what is going to reshape the business and consumer world.

Neville is the 'real deal', and it is the foolish person with a stake in this social media world who doesn't pay attention to him. Very foolish, indeed.

When I started podcasting in January 2005, anyone's expectations about audio quality didn't matter much. Unless you were a radio station with a studio, what you could expect was pretty rough and ready audio quality, especially if you wanted to record remotely.

Nearly twenty years later, the picture could not be more different.

From its humble beginnings, where enthusiasts and amateurs alike shared their voices through rudimentary setups, podcasting has undergone a remarkable transformation. Today, it stands at the forefront of digital storytelling, with studio-quality audio no longer a luxury but a baseline expectation among discerning listeners.

This seismic shift in listener expectations has been driven by a confluence of technological advancements and cultural trends. As the world becomes increasingly connected, and the barriers to content creation are lowered, listeners are no longer just seeking content that informs and entertains; they demand an auditory experience that is immersive, crystal clear, and professional. This evolution mirrors the journey of other media forms, where quality has transitioned from a differentiator to a necessity.

In this context, podcasters, whether seasoned professionals or passionate newcomers, find themselves at a pivotal juncture. The imperative to 'up their game' resonates more profoundly than ever. This is not merely about keeping pace but about embracing the tools and technologies that transform good content into great experiences.

The Key Essentials for Podcasting

As a podcaster, I consider three things as essential for creating quality content where, in this context, 'quality' means an excellent listening experience:

1 Microphone: Once you have decided on which type of microphone you want to use—USB connected directly to your computer, or XLR connected to a mixer and from there to your computer—the most important attribute of a microphone for a podcaster is its sound quality. A high-quality microphone can capture clear and crisp audio, which is essential for a professional-sounding podcast, and for the listening experience.

2 Mixer or audio interface: If you use an XLR microphone, the single most important feature of a mixer for a solo podcaster (as I am, but connecting to others remotely via Internet-based recording methods) is the ability to control and adjust the audio levels for my recordings, especially the 'gain'. This feature allows you to manage the volume and balance of different audio sources, such as your own voice, music, or sound effects, in real-time. It provides you with the necessary control to ensure a high-quality recording and a well-balanced final product.

3 Recording and editing software: The most important feature of audio-recording software for a podcaster is its ease of use and flexibility. The software should be easy to navigate and use, allowing you to focus on the content rather than the technicalities. It should offer flexibility in terms of editing, multitrack recording, and seamless integration with other hardware and software tools.

For many years, I used a **Blue Yeti**, a USB microphone that served me well. I've also acquired others such as **Audio Technica AT2020** and **RØDE NT-USB Mini**, all USB microphones. And all very good.

In August 2023, I bought a just-launched **RØDE PodMic USB**, a new generation of dynamic microphones combining both USB and XLR connectivity. Initially, I used the USB interface as that had been my experience so far. However, I quickly concluded that XLR would be the better bet for what I wished to do—to improve the quality of my audio recordings and increase my understanding and knowledge of audio editing and production.

The dual XLR and USB connectivity allows the mic to be used in various setups, from traditional studio connections to direct computer or mobile device connections, offering flexibility and major convenience. It would give me the ability to use a USB connection if I wanted to record via my laptop, smartphone or tablet.

To understand the differences between USB and XLR microphones, **RØDE** has a good explainer.

So to use XLR, I needed a mixer or audio interface that would convert the analogue audio from the XLR-connected mic to the digital output the computer requires. After a little research, I settled on the **Vocaster One** audio interface which has proven to be an outstanding part of my overall podcasting setup.

A standout feature of the **Vocaster One** is the auto-gain setting that you set via the three-bar symbol. Once pressed, you then speak into the microphone for ten seconds while the mixer works out the optimum gain for you. It's very good.

You can also set the gain this way via the **Vocaster Hub** companion app as well as enable one of four presets for voice enhancement. The Hub is actually your mixer where you can manipulate audio from multiple inputs if you desire. **Focusrite**, the maker of the **Vocaster One**, has a great video on YouTube with a detailed explainer of the **Vocaster Hub** and all its features and functions.

The **Vocaster One** is designed for audio podcasting, especially in a setup like mine: a solo podcaster. If you produce a podcast that, say, involves guests, both physically at your location and remotely connecting via the Internet or phone, and/or you live-stream a show, there are more suitable options you should consider. Worth considering are other mixers from **Focusrite**, especially their **Scarlett** studio range. Take a look, too, at RØDE's offerings, including the excellent **RØDE-caster Pro II**. And if you have a large budget, consider newer wireless products such as the **Nomono Sound Capsule** portable podcast studio.

Finally, there's the software I use for audio recording, editing and production—**Adobe Audition**, a powerful tool for audio podcasting and, in my experience, the best product of its type for Windows computers. I've been using this in its various versions for over a decade.

For one of the three podcasts I host or co-host, my prime role other than hosting is episode production. My two co-hosts and I typically record via **Zoom**. I use **Adobe Audition** to edit the three individual audio tracks, either from the Zoom recordings or from the local recording each of us makes, depending on which is the better quality and thus requires less editing or production processing. The multitrack mode in **Adobe Audition** is an essential tool for this; it is intuitive to use and quick in its processing on the highly-specified Dell XPS desktop PC I have.

Conclusion

In summary, I find the combination I've outlined is the best setup I've had as a solo podcaster + remote co-hosting with others.

In particular, I would highlight the **Vocaster One** mixer as a highly versatile and user-friendly solution for solo podcasters, offering high-quality sound, easy setup, and various features to enhance and customise audio for podcasting and streaming purposes. The auto-gain feature is brilliant, and the easy-to-reach microphone mute button is a boon!

If you're a podcaster looking to up your game by making a reasonable and sound investment in studio-level hardware (and software), at a total cost of less than £600 (roughly €700 or $750), or if you're looking to get started with a quality setup from scratch, a **RØDE PodMic USB/XLR** + **Vocaster One** + **Adobe Audition** combination would be worthwhile.

Rode PodMic USB/XLR: Summary of key features

• **Broadcast-quality sound**: The PodMic offers a rich, detailed sound optimised for podcasting, live streaming, and other speech applications.

• **Internal pop filter and shock mount**: It comes with an internal pop filter to minimise plosives and an integrated shock mount to reduce vibrations, allowing for clear and professional sound quality. It also includes a studio-quality external pop filter specifically made for this microphone that fits it like a glove.

• **Versatile connectivity**: The PodMic USB model offers both XLR and USB connectivity, making it easy to connect to an audio interface, mixer, computer, or even iOS and Android devices.

• **Robust construction**: The microphone is built with a robust all-metal construction, ensuring durability and longevity.

• **Optimized for speech applications**: The PodMic is specifically tailored for speech and broadcast, with a dynamic capsule that minimises room noise and a tight cardioid polar pattern for superior room noise rejection.

Vocaster One: Summary of key features

• **High-Quality Sound**: The Vocaster One provides broadcast-quality sound, offering studio sound in seconds from its high-quality mic input, with over 70dB of gain, suitable for any XLR mic.

• **Enhance Edit**: It allows greater control of compression, EQ, and rumble reduction for each preset, providing the user with the ability to enhance their sound with one click.

• **Vocaster Hub**: This feature allows users to set levels, enhance their sound, route audio from their phone and computer, and hear their show mix. It also works with screen readers, allowing control of Vocaster's key features.

• **Phantom Power:** If you want to use a condenser microphone that requires power to work, the Vocaster One can deliver the required 48V power.

• **Simple Setup for Podcasting and Conference Calls**: The Vocaster One can be connected to a computer through USB-C to record into podcast recording software, such as Adobe Audition, which can be used to customise sound. It also allows users to select Vocaster as the audio input on Zoom and Teams. Skype, too.

• **Complete Setup for Podcasts and Streaming**: The mixer offers various connectivity options, including a headphone output, speaker outs, phone input, and camera output, making it suitable for professional podcast audio recording and streaming.

• **Auto Gain and Enhance**: The Auto Gain feature sets the correct audio level with the click of a button, and the Enhance button brings out the best of the user's voice with a choice of four voice presets.

• **Mute Button**: It allows users to quickly mute audio to avoid interruptions.

Adobe Audition: Summary of key features

• **Multitrack, Waveform, and Spectral Display**: Adobe Audition provides a comprehensive toolset that includes multitrack, waveform, and spectral display for creating, mixing, editing, and restoring audio content.

• **Recording and Mixing Capabilities**: It allows users to record and mix live through a computer microphone and studio recording equipment, or work with recorded tracks. This is essential for podcasters who need to create live narration, edit and enhance sound, or mix several tracks dynamically.

• **Audio Mixing and Mastering**: Adobe Audition is suitable for mixing and mastering audio content for podcasts. It offers over 50 effects and analysis tools, customisable controls, and the ability to remove background noise, making it ideal for refining and polishing audio content.

• **Metadata editor**: This allows you to enter the texts for ID3 tags and add cover art directly to your output MP3 file from within the program.

• **Integration with Adobe Creative Cloud**: Audition integrates efficiently with Adobe Creative Cloud, allowing for seamless collaboration and the easy addition of mastered audio to other projects in, for example, Adobe Premiere Pro.

• **Free Trial**: Adobe offers a free trial of Audition, allowing users to explore its features and capabilities before making a purchase.

Neville Hobson

Social Strategist, Communicator, Writer, and Podcaster with a curiosity for tech and how people use it. Believer in an Internet for everyone. Early adopter (and leaver) and experimenter with social media. Occasional test pilot of shiny new objects. Avid tea drinker. nevillehobson.com

Useful links

- nevillehobson.com
- nevillehobson.com/podcasts/
- forimmediaterelease.biz/weblog/comments/the_hobson_and_holtz_report_podcast_1_january_3_2005
- soundsprofitable.com/research/the-podcast-landscape/
- logitechg.com/en-gb/products/streaming-gear/yeti-premium-usb-microphone.988-000229.html
- audio-technica.com/en-gb/at2020

- rode.com/en/microphones/usb/nt-usb-mini
- rode.com/en/microphones/usb/podmic-usb
- rode.com/en/about/news-info/xlr-vs-usb-microphones-which-is-better-for-you
- focusrite.com/products/vocaster-one
- focusrite.com/
- youtube.com/watch?v=oattMM-MheU [a detailed explainer of the Vocaster Hub and all its features and functions]
- focusrite.com/scarlett
- rode.com/en/interfaces-and-mixers/rodecaster-series/rodecaster-pro-ii
- nomono.co/products/sound-capsule
- adobe.com/uk/products/audition.html

best social media platforms for podcasts in 2024

INTRODUCTION

With the rise in popularity of podcasts, it has become increasingly important for podcasters to have a strong presence on social media. Social media platforms offer a way for podcasters to connect with their audience, promote their content and increase their reach. But with so many platforms available, which ones are the best for podcasters in 2024? In this document, we will explore the top social media platforms for podcasts in 2024 and how podcasters can use them to their advantage.

There are numerous platforms available today for both podcast creators and listeners. Here, we list them in a subjective ranking based on user experience, special features, and overall popularity.

1 Apple Podcasts: With an extensive range of podcasts and a user-friendly interface, Apple Podcasts takes top spot. It's a default choice for most iOS users and offers podcast creators great visibility. Apple Podcasts is the 400-pound gorilla in this space, owning 48 per cent of the market.

2 Spotify: Known primarily for music, Spotify has made significant strides in the podcasting space. It's a top choice for Android users and offers creators access to a large, diverse audience.

3 Google Podcasts: As a Google product, Google Podcasts offers seamless integration with other Google services. It's an excellent choice for Android users, but its current limited features for creators place it third.

4 Pocket Casts: Pocket Casts boasts a beautiful UI and extensive features for listeners. While it's less known than the giants above, it offers creators a dedicated listener base.

5 Castbox: Castbox is a newer player in the market. It has a wide selection of podcasts and unique features like in-audio search, but its smaller user base ranks it last in this list.

Please note that this ranking is subjective and the 'best' platform can vary based on individual needs and preferences.

Let's now dive deeper into the top three platforms.

Apple podcasts

Apple has long been the home for large swathes of the internet's citizenry looking for music and also looking for podcasts.

What is special about Apple's podcast service for podcast creators and podcast listeners?

Apple's podcast service, also known as Apple Podcasts or iTunes Podcasts, is a popular platform for both podcast creators and listeners. It was first introduced in 2005 and has been continuously evolving to become one of the leading sources for podcast content.

I will explain what makes Apple's podcast service so special and why it is a top choice for both podcast creators and listeners.

User-Friendly Interface

One of the key reasons why Apple's podcast service is favoured by both creators and listeners is its user-friendly interface. It allows for easy navigation, making it simple for users to find and access their desired podcasts.

For podcast creators, the interface offers a seamless process for uploading and managing their content, allowing them to focus on creating quality episodes rather than struggling with technical difficulties.

Wide Audience Reach

Apple's podcast service boasts a vast audience reach, making it an attractive platform for podcast creators. With over 1 billion active Apple devices worldwide, including iPhones, iPads and Macs, the potential audience for podcasts on this platform is substantial.

Additionally, Apple Podcasts is available in over 175 countries, making it a global platform for creators to reach a diverse audience.

Integration with Apple Ecosystem

Another unique aspect of Apple's podcast service is its integration with the wider Apple ecosystem. As most users own multiple Apple devices, they can easily access their favourite podcasts across different platforms seamlessly.

Furthermore, the integration also extends to other features such as Siri, which allows users to control and play podcasts hands-free.

Powerful Analytics for Creators

For podcast creators, having access to analytics is crucial in understanding their audience and improving their content. Apple's podcast service provides powerful analytics that offer insights into listener demographics, episode performance, and overall trends.

These analytics help creators make informed decisions on their content strategy and engage with their audience better.

Discoverability

With over 2 million podcasts currently available on Apple Podcasts, discoverability can be a challenge for creators. However, Apple's platform offers various features to aid in discoverability, such as curated collections and personalised recommendations based on listening history.

Moreover, podcast listeners can easily search for specific topics or keywords, making it easier to find new content.

Monetisation Opportunities

Apple's podcast service also allows for monetisation opportunities for creators. Through its subscription feature, listeners can financially support their favourite podcasts and access exclusive content.

Additionally, the platform offers advertising options for creators to generate revenue through sponsorships and partnerships with brands.

Final Thoughts

In conclusion, Apple's podcast service stands out from other platforms due to its user-friendly interface, wide audience reach, integration with the Apple ecosystem, powerful analytics for creators, discoverability features, and monetisation opportunities. These factors make it a top choice for both podcast creators and listeners alike. With its continuous evolution and commitment to providing quality content, it is no surprise that Apple Podcasts remains a dominant force in the world of podcasts.

Spotify

Spotify provides a robust platform for podcast creators and listeners alike. For creators, it offers tools like Spotify for Podcasters, which allows for the upload and management of podcast episodes directly on the platform. Analytics provided by Spotify give creators insights into their audience demographics, listening habits, and episode performance. For listeners, Spotify offers a vast library of podcasts across various genres. The platform's AI-driven algorithm recommends podcasts based on users' listening history, helping them discover new content that matches their interests. Additionally, listeners have the option to download episodes for offline listening, making Spotify a convenient tool for podcast enthusiasts.

Spotify's commitment to the podcast market was evident when it acquired companies like Gimlet Media and Anchor in 2019. These strategic acquisitions have allowed Spotify to expand its library of exclusive podcasts, attracting high-profile creators like Joe Rogan and Michelle Obama. This move has not only solidified Spotify's presence in the podcast industry but also sparked competition among other streaming services to invest in podcast content.

Moreover, Spotify's integration with social media platforms like Instagram and Facebook has further enhanced its reach to a wider audience. By allowing listeners to share their favorite podcasts on their social media profiles, Spotify has effectively leveraged the power of word-of-mouth marketing. This strategy not only helps in promoting shows but also encourages users to engage more actively with the platform.

Google

Google's podcast platform is one of the most popular and widely used platforms for listening to podcasts. It offers a wide range of features and benefits that make it stand out from other podcasting platforms.

One of the key aspects that sets Google's platform apart is its integration with other Google services such as Google Podcasts, Google Play Music, and Google Assistant. This allows for a seamless and convenient listening experience, as users can easily access their favourite podcasts on multiple devices through these integrated services.

Additionally, Google's platform offers a user-friendly interface that makes it easy for both creators and listeners to navigate. Podcasters can easily upload their episodes and manage their content, while listeners can discover new shows and organise their subscriptions in a simple and intuitive manner.

Another major advantage of Google's podcast platform is its discoverability. With a vast library of podcasts and advanced search algorithms, users can easily find new and relevant shows based on their interests. This also benefits podcast creators as it increases the visibility of their content to potential listeners.

Moreover, Google's platform supports both audio and video podcasts, providing a diverse range of content for users to enjoy. It also offers automatic episode downloads, offline listening, and customisable playback speed, making it a convenient and personalised experience for listeners.

Furthermore, Google's platform has a strong focus on providing accessibility for all users. It offers transcript support for episodes, allowing those with hearing impairments to still enjoy podcasts. It also has a feature that allows users to easily share episodes with their friends and followers on social media.

In summary, Google's podcast platform stands out for its seamless integration with other Google services, user-friendly interface, discoverability, diverse content options, accessibility features and personalised listening experience. These factors make it a top choice for both creators and listeners in the world of podcasting.

Facebook

As one of the largest social media platforms, Facebook is a great tool for podcasters. With over 2.8 billion monthly active users, it provides a huge potential audience for podcasters to connect with. Facebook allows podcasters to create a page specifically for their podcast, where they can share episodes, engage with followers and even run ads to reach a wider audience. Additionally, Facebook groups provide a space for podcasters to connect with like-minded individuals and promote their content.

Instagram

With its visually appealing interface and over 1.2 billion monthly active users, Instagram is another top platform for podcasters. Instagram's focus on visuals allows podcasters to create eye-catching graphics and videos to promote their episodes. The use of hashtags also makes it easier for podcasters to reach a wider audience and engage with followers. Additionally, Instagram offers the option to go live, making it a great platform for hosting Q&A sessions or behind-the-scenes content.

Twitter/X

Known for its fast-paced nature and real-time conversations, Twitter (as it has been known since its arrival in July 2006, only being renamed 'X' in 2023) is a great platform for podcasters to engage with their audience. With over 353 million monthly active users, podcasters can use Twitter/X to share updates about their show, interact with followers and join relevant conversations using hashtags. Additionally, Twitter/Xs audio feature allows podcasters to share short snippets of their episodes or engage in live-tweeting during an episode release.

YouTube

While primarily known as a video platform, YouTube also offers great opportunities for podcasters. With over 2 billion monthly active users, it allows podcasters to reach a wider audience through the use of video content. Podcasters can create visually appealing videos that showcase their episodes, and even go live to host Q&A sessions or interact with their audience. Additionally, YouTube's search engine makes it easier for new listeners to discover the podcast through SEO optimisation.

TikTok

With its short-form video format and over 1 billion monthly active users, TikTok has become a popular platform for content creators, including podcasters. Podcasters can use TikTok to create engaging and entertaining videos that promote their episodes and reach a younger audience. Additionally, TikTok's algorithm makes it easier for new users to discover content, making it a great platform for podcasters looking to expand their reach.

. . .

Best-in-class podcast file hosts

These are the best sites for hosting your podcast files, and these hosts store both your podcast mp3s and your log-ins and passwords for uploading to Apple, Spotify, Google and other podcast outlets. Once you set the login/etc., details they stay 'in the system' so you don't have to enter them every time you go to publish your latest podcast episode.

Libsyn, one of the most popular podcast hosting platforms, offers a wide range of features and benefits to content creators. It provides unlimited bandwidth, advanced analytics, scheduling capabilities, and the ability to distribute your podcast to multiple directories. In addition, Libsyn stands out with its monetization options, allowing you to generate income from your podcast through premium content or ad sponsorship. Notably, the platform also ensures that you retain full ownership of your content. However, Libsyn's user interface may seem slightly outdated and less intuitive compared to newer platforms, making it a bit challenging for beginners to navigate. The pricing could also be a deterrent for some, as there are cheaper alternatives on the market. I've been using Libsyn since I first started podcasting in 2005 and I can't see me moving to another platform, but one learns to 'never say never' in this industry.

Other podcast hosts, such as **Podbean** or **Anchor**, also have their unique pros and cons. Podbean, for instance, offers an easy-to-use interface and competitive pricing, but its analytics are not as comprehensive as Libsyn's. Anchor, on the other hand, provides free hosting, making it an attractive option for those on a budget. However, its monetization options are less versatile, and it lacks some advanced features, such as scheduling. Thus, the choice of a podcast host depends largely on your specific needs and priorities as a content creator. It's essential to consider factors like budget, ease-of-use, distribution options, and monetization potentials when making your decision.

Conclusion

In 2024, social media will continue to play a crucial role in the success of podcasts. Each platform offers unique opportunities for podcasters to connect with their audience, promote their content, and increase their reach. By utilising a combination of these top platforms, podcasters can create a strong presence online and continue to grow their listenership in the years to come.

References

• Statista. (2021). Number of monthly active Facebook users worldwide as of June 2021. Statista. https://www.statista.com/statistics/264810/number-of-monthly-active-facebook-users-worldwide/

• Instagram Press Center. (n.d.). About Instagram. Instagram. https://about.instagram.com/about-us

• Twitter Company Facts and Figures. (2021). Twitter. https://about.twitter.com/en_us/company.htm

• Statista. (2021). YouTube: technical and SEO facts. Statista. https://www.statista.com/study/19573/youtube-facts-and-statistics/

• TikTok for Business. (n.d.). About Us: Our Story. TikTok for Business. https://www.tiktok.com/business/en/about-us/our-story

mainstream media's podcasting successes are no threat to the independent podcaster - shel holtz

I AM THOROUGHLY HOOKED ON 'INVISIBILIA'.

One of the new crop of podcasts from National Public Radio (NPR), Invisibilia delves into the hidden forces that influence our lives. I listened to the first episode based on a recommendation in a blog post listing worthwhile podcasts. By the end of the episode, I was moved to tears. I had to pull my car over; driving and crying isn't safe.

Since then, I have been binge-listening. Invisibilia was broadcast weekly with podcast episodes made available after the broadcast, but the first season wrapped up in February, well before I started listening, so I was able to grab them all at once. The show is even more compelling than the first mainstream podcast hit, 'Serial', also from NPR. While Serial's weekly exploration of its miscarriage-of-justice tale was riveting, it didn't pack the emotional wallop of Invisibilia. I have also found two co-hosts (Alix Spiegel and Lulu Miller) make for more engaging listening than one.

Podcast listeners must agree. Back in February 2023, according to Poynter, 'Invisibilia' replaced 'Serial' as the most popular podcast in iTunes' rankings.

The production values in 'Invisibilia' are stellar—which you would expect for a program that also airs on 260 public radio stations across the U.S. Many of NPR's podcasts, though, are original podcast-only content—including Serial—and NPR and its affiliates are clearly investing in these productions, too. NPR is not alone. Other media properties, like Slate, are also producing podcasts with the help of professional editors and dedicated staff.

For an independent podcaster like me, the prospect of competing with these programs for audience's share of ear is daunting. One podcaster warned that the success of shows like Serial and Inivisibilia "are reserved for the few elite media properties that have a vital component: a built-in audience. Without that springboard, your show is a hobby with a tiny listenership."

When mainstream media first embraced podcasting, it was strictly a means of expanding the audience for broadcast programming. Here in the Bay Area, for example, the local CBS radio affiliate includes a 3-4-minute chat with football legend John Madden each weekday during a morning sports report. At the conclusion of the segment, the audience is reminded of the two times each day they can tune in to hear Madden's observations; they can also subscribe to the podcast.

There's no question that podcasting is on the rise. Forty-seven million Americans over the age of 12 listen to a podcast at least once a month; in total Americans consume 21 million hours of podcasts every day, according to Edison Research. This growth has been incremental since podcasting's birth. Serial and Invisibilia's success have boosted awareness, but don't account for the continued growth in hours consumed, number of people listening, and number of shows available.

The question is whether mainstream media's sudden focus on top-tier original content threatens the independent podcaster. Will the high production values of Invisibilia keep listeners from spending time with shows produced in someone's home office in their spare time?

I'm not worried.

Mainstream media has recognized the growing popularity of time-shifted audio and the ability to deliver general-interest programs to sizable audiences without needing to convince hundreds of local radio stations to air the show. The fact that general-interest shows are finding an audience as podcasts can only be good for the independent podcaster as audiences begin scouring directories for new titles.

One key difference between shows from NPR, CBS, and other traditional media and independent podcasters (besides the budgets to crank out high-end audio) is the audience. While much of the content from big networks has broad appeal, independent podcasters tend to target niches. That was podcasting's promise from the very beginning: You can produce a show for next to nothing that attracts a small target audience, the kind of show no radio station would ever agree to air since the audience is too small.

How small the audiences of these targeted podcasts remain depends how well the content serves the needs of its niche and how good the podcaster is at promoting the show. While some podcasters may complain that, without a built-in audience, they are doomed to "tiny listenership," nobody ever made the same dire prediction about blogs. Bloggers with no name recognition have attracted millions of readers. So have hosts of YouTube shows.

There is no logical reason to assume podcasters can't achieve the same kind of success.

Podcasting is not like Broadway, where every production is judged against every other show in town. It's more like television, which accommodates big-budget scripted dramas alongside low-budget reality shows and no-budget public-access programming. There's room for all kinds of content in the world of podcasting, and each can find its own level. Consider that mainstream podcasts are commanding some of the highest advertising rates anywhere—from $25 to $45 CPM—for shows with at least 20,000 downloads per episode. At the same time, shows with significantly fewer listeners are inking different advertising deals with companies seeking to reach their very narrowly defined audiences.

Besides, according to the Edison report, people who listen to podcasts listen to more podcasts than any other kind of audio content, including their own music collections, AM/FM radio, and streaming media. If a podcast fan finds a new podcast she likes, she'll just add it to the queue, and those 21 million hours of podcasts heard every day will grow to 30, 40, 50 million.

Invisibilia and Serial are the proverbial tides that can only lift all boats. With luck, that will include the FIR Podcast Network, which launches a new website within the next week or so in the hopes of attracting more listeners from within the organizational communications niche.

When you're done listening to Invisibilia, give one of our FIR shows a try. Our production values don't come close, but for communicators, our content is gold.

Update

A number of news-focused mainstream podcasts are biting the dust. The reason? They cost too much to produce and the cost isn't recouped by the ad revenue. We're talking about the shows that send crews around the world to record interviews and events, with high levels of production values. Think Serial. The shows being canceled include one of my favorites, Invisibilia. There has been a fair amount of coverage of this, including an episode of "On the Media," which featured Libsyn's Rob Walch (who interviewed Neville and me back in the day for his long-gone "Podcast 411" podcast).

In addition to his business communication and consulting and development work, Shel is a regular speaker on topics surrounding the application of online technology to strategic organizational communication. He speaks regularly at IABC and Ragan Communication conferences. He also has been on programs for the International Quality and Productivity Center, The American Association of Collegiate Schools of Business, the American Gas Association, the American Institute of Certified Public Accounts, the Association of California Water Agencies, and the Western Pension & Benefits Conference, among others. He also spoke regularly on the Lexis-Nexis touring seminar, "Communicating in a Wired World."

Shel is a five-time winner of IABC's Gold Quill award (recently winning for external Web sites) and was named IABC/Los Angeles's Communicator of the Year. Shel served six years on IABC's executive board, two as director of U.S. District 6 and four as a director-at-large. While a director-at-large, he held the Technology portfolio, among others. During his term on the board, he served three times as a member of the executive committee. He has also served as chapter president for IABC/Los Angeles and in a variety of other volunteer positions in IABC at the chapter, district and international levels. Shel earned his IABC accreditation (ABC/Accredited Business Communicator) in 1984.

In 2005, Shel was named an IABC Fellow, the highest honor IABC confers on its members. He also is an Accredited Business Communicator through IABC.

Shel and Neville Hobson co-host the influential podcast 'For Immediate Release', which is aimed at business communication professionals (think Marketing, PR, Public Speaking, Content Creation, Social Media Management, and Corporate-level Business Communication).

shelholtz.com

firpodcastnetwork.com/for-immediate-release

should i script my podcast? - donna papacosta

ONE OF THE most popular questions I hear from novice and would-be podcasters is: “Should I script my podcast?”

Queries related to podcast scripting are also the most common keyword searches on my blog.

I’ve written this chapter to help answer these questions, so you can benefit from my many mistakes. (Lesson in humility: listen to your first five podcast episodes.)

Of course, the answer to the podcasting scripting question is not a straight “yes” or “no.” It’s more like: “It depends.”

In this chapter we’ll explore:

- When you should script your podcast.
- When you should not script your podcast.
- How to write an effective podcast script.
- How to read a podcast script naturally.
- What an effective podcast script might look like.

. . .

Let's start with the positives and negatives of scripting.

Conversational podcasts don't need a script

If your podcast consists of interviews or conversations between you and one or more guests, I'd suggest not using a script at all. Instead, write out a list of topics you want to discuss with your guest, and share this list with him. However, do not send the precise questions you're planning to ask. Why not? Because some guests, believe it or not, will scribble out their exact answers and then expect to read them verbatim during the recording session. This is especially true when someone is nervous and has little experience with interviews.

In my early days of podcasting, I wanted to be nice, so I would send people questions in advance when they asked for them. As a result, I had to stop people dead cold in the middle of one of their answers to say: "Wait a minute. You sound a bit stilted. Are you reading? Please don't. Let's have a conversation instead. You'll thank me later." And they did.

Needless to say, I no longer email exact questions to guests in advance – just general question areas.

Why am I so against this type of reading? Unless you're a trained voice actor, you're probably not very convincing reading your answers. In fact, it's likely that you'll read in a monotone and bore people to death. Trust me: They can tell you are reading.

The structure of your language and the tone of your voice are giveaways.

As the podcast host, if you are afraid of forgetting something, by all means write your main points down as a checklist for yourself. This includes the questions for your podcast guest. Keep in mind that your podcast will most likely be edited. In the worst case, you can record and add forgotten content later. So, don't worry about leaving something out.

In my experience, the best interviews happen when I start off with questions that are merely guideposts, and let the conversation branch off naturally, with an easy flow. Don't worry if you ask only five of your seven planned questions, as long as the content is good, and your listeners will benefit from hearing the conversation.

Interview tips for podcasts

If you are recording your podcast interview in person, you have a bit of an advantage, I think. You and your guest can see each other and read each other's body language. Most podcast interviews, however, are recorded remotely. (I use Skype, Zoom and Call Recorder on my MacBook Pro most of the time.)

Let's imagine you are in Chicago and your guest is in Paris. You have sent her question areas, not exact questions in advance, and she knows what the goals of the interview are. Before you start recording, try to set her at ease. Sometimes a podcast guest is quite jittery at the start, and you'll do everyone a favour if you can help her relax. Believe it or not, I've interviewed CEOs who were quaking in their boots.

To counter the nerves, chat about the weather in Paris. Ask about her kids or her dogs or her latest project. Don't talk too much about the content you're going to discuss in the podcast, however, because you want it to be fresh while you're recording.

Before you begin the formal interview, remind your guest that you won't be uttering agreements as she talks. In other words, you will avoid interjecting such phrases as "Oh, I see" or "Uh huh" while she speaks, as you normally would during an in- person conversation. (Well, I normally would. One of the drawbacks of growing up in New York is my tendency to want to jump in when someone speaks. I don't like to think of this New York trait as interrupting per se; I'm adding value!)

Also, do tell your guest you will leave a pause when you think she has finished saying something. By the same token, you want her to pause when you have finished speaking.

Without the visual cues, it's hard sometimes to know when someone is actually done making a point, or if they're just pausing momentarily for emphasis or to gather their thoughts. You want to avoid your voices jumping on top of each other. It's simple to edit out dead air. It's harder to edit two voices laid on top of each other.

For the interviewee

Since 2005 I've conducted hundreds of podcast interviews, many in person but most remotely. Various factors affect the success of a podcast, both from a content and audio quality point of view. But you, as the interviewee, are the most important.

Here are a few tips to help you shine as a podcast guest. After all, you're probably doing the interview to share your knowledge as an expert, to promote your business or your book, or to generate speaking gigs. So, you want to be your best.

Preparation

Check out the podcast to ensure it fits your needs. Ask the host what the angle of the interview is, and about the intended audience. If your target market is mega- corporations, and the podcast is geared toward home business, you should probably decline.

Ask for discussion areas in advance, but don't expect to get every question in writing. As a podcaster, I don't share exact questions before the interview, because some guests get a little nervous and script their answers, and then expect to read them on the podcast. Trust me on this: Unless you're a trained voice actor, reading can be the fastest way to ruin an interview. Of course, you do want to keep relevant facts and figures at your fingertips so you sound like the expert you are.

Get the technology out of the way early. When you're booked as a guest, confirm the date and time (including time zone). Ask whether the interview will be conducted in person, by phone or via Skype or Zoom.

Sounding great

These days, the majority of podcasters will use Skype or Zoom for remote interviews. Although the popularity of both apps continues to grow, some guests are not familiar with them, and attempt to “try it out” at the time of the interview. This is not a good idea. If you’re not accustomed to using Skype or Zoom, install it in advance and experiment with a friend or family member. Or use the “test call” feature within the application. Learn how to start up the program, plug in your headset and/or microphone, and turn off notifications so you don’t hear annoying beeps during the interview.

The right gear will improve sound quality immensely. If you’re going to do multiple remote interviews, invest in a headset and microphone combo. Even better, a USB microphone would be a step above a one-piece headset/mic combo. (The RØDE NT-USB Mini gets high ratings.) If you don’t have proper headphones, you can use the earbuds from your smartphone. Please avoid using the speakers on your computer for output, and the *worst* audio quality will result from using the microphone built into your PC. If you must rely on it, learn to use it properly and position yourself for optimal sound.

Quiet on the set! If you have a thunderous fan or air conditioner in the room, turn it off for the interview. If you have a noisy child or pet, do not do the interview if they’re within earshot, unless they’re part of the story.

Share your stories

Using a story to make a point can be very effective. But do be sure your stories don’t run too long. Check with the podcaster: how much time do you have? Remember not to talk too fast. Sometimes when we’re excited about a topic, we speak too quickly. Slow down, and remember to breathe.

Pauses please. The interviewer might ask for pauses between questions and answers, which makes the editing process easier. So, after you say something, don’t be afraid of the silence, which will be edited out. When I do interviews, I usually warn the guest in advance about these pregnant pauses.

Yes, sometimes you must script

Many times I've talked a client out of scripting a podcast for the reasons I've outlined in this book, but sometimes you have to bite the bullet and let the scripting begin.

Fellow podcaster Steve Lubetkin, commenting on a blog post of mine about podcast scripting says: "We script intros and outros, and for some clients in heavily regulated industries, we prepare scripts in advance for legal review. It's a lot easier to get lawyers to approve a script in advance and stick to it than to have them picking apart the recording afterward and having to re-assemble the guests and re-record the interviews."

I couldn't agree more. I've been in this situation with clients' podcasts too. However, I do encourage the people involved in the podcast to not sound like they're reading. They usually accept my advice when I explain that we'll probably put people to sleep if we have three or four people around a table who sound like they're reading on the podcast.

Another of my clients in the financial industry takes a different approach. We make sure the questions I'm going to ask the guest are approved in advance. We then do several takes on the answers. Some guests get better as they warm up. Then I get a transcript of the interview and send it to the client. The legal and compliance people review it to make sure we're not going to get in hot water, and then I edit accordingly. Fortunately, we've become so adept at writing the questions that the lawyers rarely ask for edits.

If clients do request edits after the interview, you have to remind them that an audio file is not like a Word document. Rarely can you just clip out a word here and there. As Steve Lubetkin says, you really don't want to have to re-assemble the guests and re-record, but this could happen if the guests have ventured into territory that makes the lawyers nervous.

> "The best interviews happen when I start off with questions that are merely guideposts, and let the conversation branch off naturally, with an easy flow."
>
> *Donna Papacosta*

Mostly retired now, Donna Papacosta is a writer, speaker, podcaster and consultant who has helped people and organizations to share their stories and build their businesses.

In September 2005 she started one of the first business podcasts in Canada, Trafcom News (ask her about how to name a podcast; learn from her mistake!) Since then she has produced hundreds of podcast episodes and has taught many organisations and individuals how to use audio, multimedia and social media effectively.

Donna's book on podcasting can be bought as a Kindle-friendly book on Amazon: https://amzn.to/48j2oc9

{Lee: Involved with social media since its earliest days, Donna has been and remains a much-loved and much-respected social media and business communication specialist and commentator. I've met Donna face-to-face at a conference and she is every bit as warm and charming in person as she is on her podcast.}

the power of storytelling - donna papacosta

SUPPOSE you want to tell great stories with audio, rather than conduct interviews. The best communicators share stories, not just facts. Watch just about any TED Talk and notice how the speaker weaves a tale, using rich vocabulary and honest emotion to draw you in and inspire you.

Every podcast you produce may not be a full-blown story, but do consider adding key elements of storytelling to your show.

Remember your audience is listening, not reading.

Be sure to open with something strong to draw in listeners and give them a reason to play the whole episode. Perhaps start with a compelling question, something like: "Do you know the top three reasons why people hate your website?"

Use the "you" word. This can be the most powerful word in your podcast. Remember: you're speaking with one person at a time, not a big crowd, despite your audience stats!

Bring emotion into your story. How did the principal character feel? Whether it's an employee in your organization, a client or a customer, or you you're talking about, don't be afraid to incorporate emotion. Remember: If we don't feel, we don't act.

Use repetition to your advantage. It's OK to say something more than once.

Read well-written prose, which will help you to become a better storyteller. Some of my favourite storytelling writers include David Sedaris, Nora Ephron and Anne Tyler.

> "The best communicators share stories, not just facts."

Mostly retired now, Donna Papacosta is a writer, speaker, podcaster and consultant who has helped people and organizations to share their stories and build their businesses.

In September 2005 she started one of the first business podcasts in Canada, Trafcom News (ask her about how to name a podcast; learn from her mistake!) Since then she has produced hundreds of podcast episodes and has taught many organisations and individuals how to use audio, multimedia and social media effectively.

Donna's book on podcasting can be bought as a Kindle-friendly book on Amazon: https://amzn.to/48j2oc9

{Lee: Involved with social media since its earliest days, Donna has been and remains a much-loved and much-respected social media and business communication specialist and commentator. I've met Donna face-to-face at a conference and she is every bit as warm and charming in person as she is on her podcast.}

outline of a typical podcast script - donna papacosta

YOUR PODCAST SHOULD REFLECT you and your goals. If you're just getting started in podcasting, you may be looking for ideas for structuring your show and your script, if you're using one. Here are some ideas.

Welcome and introduction to the show

This explains quickly what the show in general is about. Don't be too repetitive here. Some podcasters repeat their name and the show name three or four times. That's too much! And please, don't tell us several times how "excited" you are about today's show. This gets old quickly.

Introduction of the episode topic

What's the benefit to your audience? Be short and sweet here and give us a teaser so we'll keep listening.

Topic or interview

If it's just you talking, be sure to outline your main points so you don't forget something important. If you have a guest on your show, introduce him graciously and generously, and point out his credentials. Then get right into the meat of the content.

Wrap-up

Because your listeners don't have visual cues, it's helpful to wrap up the podcast with a recap of main points. Do you have a call to action? Mention it now.

Comments

I suggest you save promos, comments, "housekeeping items" and other such content for the end. Don't make me listen to your podcast for 10 minutes before I derive some value from it. Be sure to tell listeners where to reach you via the Web, email, phone, etc.

> "Don't make me listen to your podcast for 10 minutes before I derive some value from it."

Mostly retired now, Donna Papacosta is a writer, speaker, podcaster and consultant who has helped people and organizations to share their stories and build their businesses.

In September 2005 she started one of the first business podcasts in Canada, Trafcom News (ask her about how to name a podcast; learn from her mistake!) Since then she has produced hundreds of podcast episodes and has taught many organisations and individuals how to use audio, multimedia and social media effectively.

Donna's book on podcasting can be bought as a Kindle-friendly book on Amazon: https://amzn.to/48j2oc9

{Lee: Involved with social media since its earliest days, Donna has been and remains a much-loved and much-respected social media and business communication specialist and commentator. I've met Donna face-to-face at a conference and she is every bit as warm and charming in person as she is on her podcast.}

how to write a podcast script - donna papacosta

WHY DO I discourage scripting and reading a podcast? First, most people are pretty lousy script readers, as I've shared with you in a previous chapter.

Secondly, I think the best podcasts sound conversational rather than scripted. That being said, it is also true that if you can fake a conversational tone, you might do all right with a script. Faking entails inserting the occasional pause, um, ah, and even a flub or two. You should also be sure to vary the speed of your read.

Do you think I'm disingenuous recommending this faking tactic? Well, let's suppose you've read and agree with the advice in a previous chapter, and yet you have good reasons for writing and delivering a podcast script. For example, in some tightly wound organizations, every word uttered publicly must be approved, so it's deemed easier to get permission before recording rather than after. For some of us, this would be a last resort, but it may be necessary at times.

The main point is this: Please do not pick up a written document, which is crafted for the eye, and just read it cold. I can almost guarantee this won't work. You need material that is written for the ear.

Writing for the ear

Sometimes people forget this simple truth: When we listen to what you are saying, it's not the same as reading what you have written. Copy that looks fine on the page or screen does not always sound right to the ear. Worse, it can be hard to understand. The listener does not have visual cues. In other words, there are no headlines, subheads, bulleted lists, sidebars or even paragraph breaks for him to see.

Take a typical written passage, adapted from copy on a municipal economic development website:

> With easy access to North America's third largest financial centre and a workforce armed with expert knowledge in a variety of fields, establishing your business in Hooterville will pay handsome dividends. Located at the epicenter of Canada's Golden Horseshoe, Hooterville, a dynamic community of 150,000 residents, is well within reach of major U.S. capital markets and nearly seven million potential consumers in southern Ontario. Coupled with a favourable Canadian tax environment, Hooterville makes perfect business sense.

This is fairly well written, except for the dangling participle in the first sentence. Let's now recast it for the spoken word:

> Think about establishing your business in Hooterville. Here's why: Hooterville offers a talented workforce and easy access to North America's third-largest financial centre – Toronto. In fact, Hooterville is at the heart of Canada's Golden Horseshoe, with a dynamic community of 150,000 people. What's more, the town is close to major U.S. markets and almost seven million consumers in southern Ontario. Add a favourable Canadian tax environment, and Hooterville makes perfect business sense.

Do you see the difference? Shorter sentences, simpler language, and a more direct tone.

Writing tips

Here are a few tips for writing for the ear, whether your end product is a script, speech or podcast:

- Use simple words, not complex ones. For example: "Use" rather than "utilize."
- Shorten your sentences. If it requires a semicolon, it's probably too long.
- Round all numbers. Say "nearly one million," not "989,320," unless there's a reason to use the exact figure.
- Use the active voice, not passive. "Our team ran the webinar," not "The webinar was run by our team."
- Use less formal language. For example, use contractions, as long as you can enunciate clearly. (Won't rather than will not.) Be sure listeners can hear the difference between your pronunciation of can and can't.
- Give auditory guideposts. Say things like: "Let's talk about three ways to use social media to market an event. First..." Add transitions between each of your points, and a recap at the end, using your numbered list as a structure.
- Don't be afraid of repetition. It's OK to repeat important information for emphasis.

> "Copy that looks fine on the page or screen does not always sound right to the ear ... The listener has no visual cues ..."

Mostly retired now, Donna Papacosta is a writer, speaker, podcaster and consultant who has helped people and organizations to share their stories and build their businesses.

In September 2005 she started one of the first business podcasts in Canada, Trafcom News (ask her about how to name a podcast; learn from her mistake!) Since then she has produced hundreds of podcast episodes and has taught many organisations and individuals how to use audio, multimedia and social media effectively.

Donna's book on podcasting can be bought as a Kindle-friendly book on Amazon: https://amzn.to/48j2oc9

{Lee: Involved with social media since its earliest days, Donna has been and remains a much-loved and much-respected social media and business communication specialist and commentator. I've met Donna face-to-face at a conference and she is every bit as warm and charming in person as she is on her podcast.}

steve lubetkin's podcast journey

I GOT bit by the radio bug in high school. My dad worked at Fort Monmouth, the military base in central New Jersey where the US Army had its training school for the Signal Corps, the unit that handled the Army's communications needs. The base taught photography, television, and radio skills to military and civilian personnel from the US Armed Forces and from our allies.

When I was about 12 or 13, my dad arranged for me to spend an afternoon at the mock radio studio where personnel trained to become Armed Forces Radio Service announcers and disk jockeys. One of the instructors taught me how to work the board, cue up vinyl records, and segue between them, interspersing station IDs and promos from tape cartridges. From that day on, I knew I wanted to be in radio.

At home, I created a studio of sorts in my parents' basement, recording make-believe radio shows using a reel-to-reel tape deck, a record turntable, and a Shure microphone. I played them back solely for the enjoyment of my best friend and me.

When I got to college, the first place I headed was the campus radio station, which agreed to try me out on the air—on the understanding that I would very quickly study for, and obtain, the FCC's Third Class Commercial RadioTelephone Operator License with basic broadcast endorsement, which would enable me to legally sign the radio station's transmitter log as the operator on duty.

I did my first full air shift at WMCX-FM in September, 1974. It was an amazing experience, getting to pick the music, running the board, talking to that unseen audience.

The following year, the station's music director, Lee Mrowicki, a club deejay who had strong connections in the Asbury Park music scene, had landed himself a position voicing commercials at WJLK, a major station at the Jersey Shore owned by the local newspaper. Lee learned they needed a licensed broadcast engineer to produce public affairs shows on Sunday nights. He recommended me for that position, which mainly involved playing taped shows over the air, one after another, from 6-9 on Sunday nights. After a while, a weekend newscaster position opened up, and I started working in radio news on Friday and Saturday nights.

From there I started doing fill-in shifts in news and production and learned a lot about commercial radio.

I stayed in radio for almost five years—including a memorable assignment in 1977, when I and another reporter became the first journalists to cover a rock concert with a portable computer. It was a Grateful Dead concert in Englishtown, NJ, and the Asbury Park Press newspaper sent us with a "portable" data terminal to file our stories.

While writing for the paper, I was also collecting audio interviews and sound from the concert that I fed back to the radio station that the Press owned.

However, after I graduated from college, I needed to make a bit more money than the part-time radio gig was paying. I bounced between a retail sales job and a radio shift at a different station, but finally took a job in print journalism—that lasted just nine months until I was recruited into a public relations position with a company that I covered as a reporter. I stayed in corporate PR positions for the next 25 years.

I tried to apply my broadcasting skills to the PR world, often using a portable cassette recorder to tape company events and then try to feed audio over the phone to the local stations in New Jersey. Today, with digital recording, websites, and social media, it is so much easier to distribute audio clips like this.

In the 1990s, I was working for a financial services firm, and we launched a telephone conference call series to promote the expertise of our analysts. I took the role as program host, and we formatted it like a radio talk show. The analysts loved it, and actually scheduled the calls around my availability so they could use me as the on-air moderator.

My corporate PR career came to an end in 2004, after the commercial bank for which I was doing PR was acquired first by one bank, and then that bank acquired by a larger one. After a year mostly spent in limbo, the new owner reorganized, and I ended up on the outside looking in. It turned out to be the best thing that could have happened to me, because that's the moment I decided to set up my consultancy.

This was right around the time that podcasting was becoming visible as the newest form of what we then called "new media." When I listened to podcasts I was struck by how cool it would be to distribute those "make believe" radio shows I was doing as a teenager. I also realized that many of the programs I was hearing were being done by amateurs who didn't know how to announce, edit, or even manage volume levels in a recording. I decided to become proficient at digital recording so that I could produce podcasts. I wasn't so interested in recording myself talking as I was in selling this service to businesses. I knew that to sell this to corporations, a podcast had to sound as professional as anything you might hear on National Public Radio or the CBC.

That's when I began acquiring digital recording gear and software, and reaching out to prospective clients looking for this kind of service.

Steve Lubetkin is a foundational member of the early adopters of social media in business. Steve and Donna Papacosta (see previous chapter) wrote the much respected and influential book, 'The Business of Podcasting: How to Take Your Podcasting Passion from the Personal to the Professional'.

. . .

Lubetkin Media Companies LLC

Telling Stories in Sound and Images

Steve Lubetkin

Managing Partner

708 Crestbrook Avenue, Cherry Hill, NJ 08003-1502

+1 (856) 751-5491 Office

+1 (856) 625-5502 Mobile

steve@lmediacos.com

@PodcastSteve@newsie.social on Mastodon

www.BeingTheMedia.com

My free/busy calendar: freebusy.link/podcaststeve

NJ Notary Public, Commission #50035289, exp. 3/30/2026

video podcasting

Video Podcasting: What You Need to Know

VIDEO PODCASTING, also known as vodcasting or vidcasting, is a form of podcasting that involves the use of video instead of just audio. It allows for more engaging content by incorporating visual media such as videos, images and slideshows into the podcast.

How Popular is Video Podcasting?

According to a survey conducted by Edison Research, 75% of Americans are familiar with podcasting and 50% have listened to a podcast. Video podcasts have also gained popularity in recent years, with more than half of podcast listeners consuming video content as well.

Best Platforms for Your Business

There are various platforms that you can use to host and distribute your video podcasts. Some popular options include **YouTube**, **Vimeo**, and **Blip.TV**. These platforms not only allow you to reach a larger audience but also provide analytics and monetization options for your video content.

Equipment Needed for Creating Video Podcasts

To create high-quality video podcasts, you will need some basic equipment such as:

• Camera: This could be a DSLR or a smartphone camera with good video recording capabilities.

• Microphone: A good quality microphone is essential to ensure clear audio in your podcasts. I use a RØDE NT1 on a desktop stand. It comes with a pop filter (essential) and has a beautiful clear tone. There's also RØDE's DSLR-mounted microphone, the VideoMic Pro+, which packs more punch than might be expected.

• Lighting: Proper lighting can greatly enhance the visual appeal of your video podcast. You can use natural light or invest in a basic lighting kit.

• Video Editing Software: This will allow you to edit and polish your footage before publishing it. I use **Adobe**'s leading-edge editor, 'Premiere Pro' but there are others around with similar feature sets. 'Premiere Pro' is part of Adobe's 'Creative Cloud' suite of programmes (aka 'apps') and is, as a standalone, expensive. But if your business subscribes to the whole Creative Cloud offering, which includes Photoshop, Audition (industry-leading audio editor), and twenty other powerful apps for creatives, you save a considerable amount, and have best-in-class apps for your creative teams to create magic with.

Tips for Creating Successful Video Podcasts

• Plan your content: Decide on a specific theme or topic for each episode and plan out the flow of your podcast.

• Keep it visually engaging: Use a mix of videos, images, and slides to add visual interest to your podcasts.

• Be authentic: Viewers respond well to natural and genuine content. Don't be afraid to show your personality and allow for spontaneous moments in your podcasts.

• Promote on social media: Use social media platforms to promote your video podcasts and reach a wider audience.

• Consistency is key: Stick to a regular publishing schedule so that your audience knows when to expect new episodes. This will also help you maintain momentum and grow your viewership.

In Conclusion

Video podcasting is a popular and effective way to engage with your audience through visual media. With the right equipment, platform, and content strategy, you can create successful video podcasts that will help grow your business and connect with your viewers on a deeper level.

Overall, there are many benefits to incorporating video podcasting into your content strategy. It allows for more dynamic and engaging content, helps reach a wider audience through various platforms, and provides opportunities for monetization and analytics. Try it and see how video podcasting can elevate your business. Remember to plan, be consistent, and most importantly, have fun creating your video podcasts!

ten rules of social media strategy

USE these 10 rules of social media strategy to ensure success when rolling out social media tools inside the organisation.

by Victoria Mellor, CEO, Melcrum Publishing

Before you jump into social media, ask yourself whether your organisation is ready. It's the job of the communicator to recognise and understand where and how social media tools might work in your business. So, here's a 10-step checklist to help ensure your social media development goes according to plan:

Assess your readiness

Before doing anything, assess your organisation's cultural readiness. Will blogs, podcasts, wikis and so on survive in your organisation? Do the new tools complement your organisation's processes and vision? Remember, social media won't necessarily suit all organisations and employees.

Focus on people, not the technology

Don't overlook the needs and preferences of the people who will use the new tools—focus on the human need for the technology and how they're able to adapt to it.

Carefully consider the business benefit

Before rushing in, think about the purpose of the tools. Once you're clear on the actual business purpose of the tools, you'll be led by the goals you're trying to achieve, rather than the technology itself. Ask specific questions, such as:

1. Who will use these tools?
2. How can they support or generate useful conversations?
3. How can they give employees better access to leadership?
4. How can they help global teams collaborate and communicate better?
5. How can they change the culture?

Grasp the difference between traditional and social media

Make sure you understand the difference between the two types of media. Without grasping the difference between the two fundamentally different methods of communication, it's unlikely you'll grasp the potential of social media. As a communicator you provide the tools, you shape their use and you set guidelines for use. The audience then decides what's useful and what's not.

Prepare to relinquish control and share the process

Social media places all users on an equal footing, so control is reduced for all organisations and leaders. Employees become equal partners in the communication process and are invited to take part in a conversation and share information, rather than have information pushed at them. Also, get comfortable with ambiguity. Simple person-to-person conversations can turn into large knowledge-sharing exercises if they're allowed to grow in whatever way they need to.

Be experimental and involve employees

Because social media isn't costly to implement, you can afford to experiment. Involving employees in trial and development encourages acceptance of the tools. It helps familiarise them with the tools so they feel comfortable using them, in their own time. Think about producing "beta" sites for employees to experiment with—this will help iron out glitches before going "live".

Give employees clear guidelines

However informal, there's no reason why communication taking place via a blog or a wiki should stray from the usual behaviours expected in any workplace interaction—such as professionalism or respect. Spell out the consequences of unacceptable behaviour. Think about prohibiting anonymity, so employees will realise they'll be held accountable for their comments. You could also outline a Social Media Code of Conduct (bit.ly/3T5qWAv).

Take a hands-off approach

People are instinctively fearful of radical change, so resist the urge to market tools aggressively. Social media tools seem to work best when they're allowed to develop and grow organically. Put together platforms and software and then consult employees about how to optimise their use and potential.

Work with what you've got

Use existing technology wherever possible, learning through trial and error, and develop tools based on what's working and what isn't. Integrate new tools and applications with your existing channels and media, rather than replacing your existing platforms.

Don’t obsess about the numbers

For most organisations using social media tools, it's too early to measure their financial success with reliability or accuracy. Given that social media tools are cheap to implement, our focus should be on what can be gained from making them available, rather than the return on investment in financial terms. Focus on the outcomes rather than the activity surrounding them (e.g., the number of comments on a blog). Also, ask for anecdotal feedback from employees.

what does chatgpt think about ai for communicators? - shel holtz

SHEL HOLTZ: If you've been listening to FIR at all, you know what Neville and I think about ChatGPT and generative AI in general. What does ChatGPT think? You'll find out in this midweek episode, but not a short one, of 'For Immediate Release', another fine podcast from the FIR Podcast Network.

Heidi Miller: This is 'For Immediate Release', the podcast for communicators.

Shel: Hi everybody, and welcome to episode number 375 of For Immediate Release. This is Shel Holtz. And no, there's no Neville this episode. This is the week between Christmas and New Year's, and Neville and Laura are away, and I thought I would take advantage of this downtime to do something I've been wanting to do for a while, and that is conduct an interview with Chat GPT.

Of course, now you're able to engage in a vocal conversation with Chat GPT, so that's what I've done. Neville and I have been talking about AI generative AI in particular for about a year now, and we've been raising a lot of issues, and I've been curious what ChatGPT thinks of all of these matters, so I've taken these questions directly to the source.

Now, I do want to share a couple of notes before I jump into the interview. First, is that this is not an original idea, the idea of having an interview with ChatGPT , I wish it were, I wish I had come up with this on my own. But somebody, I want to say it was Donna Papacosta, alerted me to a faith-based podcast called Nomad, I believe out of the UK, in which one of the hosts interviewed ChatGPT .

Interestingly, this was before it had the voice conversation capability. He used some other mechanism to convert ChatGPT 's answers into a voice. I didn't do that. I actually used the voice functionality that's built into the mobile app. But I did run into one significant challenge. I took the better part of an afternoon trying to resolve it and couldn't.

And that is feeding the audio from the app into an audio device. I tried several different things. I tried Taking the data directly through a cable from the data port on the phone into an audio device and that didn't work. It continued to come out of the speakerphone. Everything else that was internal audio on the phone played through the audio device.

I'm talking about YouTube audio podcasts, music, you name it, it behaved the way it was supposed to, but ChatGPT would only play through the speakerphone. I also tried using a native audio recorder on my phone. Same thing. Everything else recorded the way it was supposed to, even phone calls, but ChatGPT just wanted to play through the speakerphone.

So, what I ended up doing was putting my phone in a bracket on a tripod and setting it right next to the microphone, the very microphone I'm talking to right now. So, the audio that you're going to hear from ChatGPT is not quite as good as I was hoping it would be. It's certainly listenable. But it's clear that you're listening to the speaker phone being recorded through the microphone.

Second thing I wanted to point out is that I gave ChatGPT a considerable list of instructions before we started the interview. I told ChatGPT who I am. I explained the nature of For Immediate Release, the fact that we have been talking about generative AI since ChatGPT 3. 5 was released in November of 2022, and that we are focused on public relations and organizational communications and that I wanted to ask a lot of questions around that and that I was hoping to have a conversation that would sound good and right in a podcast. So, what you're about to hear is the result of that. And afterwards I'll be back with a couple of quick observations.

And by the way I am cross posting this episode of FIR to 'FIR Interviews'. No reason that you should have to listen to both. It's exactly the same file. So here is my interview with ChatGPT .

Shel: Let's start with some of the fundamentals. You were trained on a lot of various types of documents. When you're prompted to help with a PR or communications task, what type of material from your training set would come into play?

ChatGPT : When I'm asked to assist with a PR or communications task, I utilize a variety of materials from my training set. This includes public relations strategies, communication theories, press releases, media relations content, crisis communication plans, and digital marketing trends. I also draw from real world case studies and articles about successful PR campaigns and industry best practices.

This wide range of sources helps me provide informed and nuanced advice for crafting messages, engaging audiences, and navigating the ever-evolving digital landscape in public relations.

Shel: What are some of the more common uses to which you're being put in support of public relations and communications?

ChatGPT : In public relations and communications, I'm commonly used for several tasks. These include crafting and editing press releases, generating ideas for content, aiding in crisis communication planning, monitoring brand reputation, and suggesting communication strategies.

Shel: A lot of the criticism about using you to write, say, press releases is that that process loses the human element.

But most press releases written by humans don't seem to include much of the humanity that people lament is missing from an AI generated release. Do you think a press release that you generate is good enough?

ChatGPT : You make a valid point about the human element in press releases. While it's true that many press releases tend to be formulaic, and might not always reflect deep personal touch, AI generated content, including press releases, can indeed be quite effective.

I strive to produce clear, concise, and relevant content that aligns with the established norms and formats of press releases. However, the best approach might be a blend of both AI and human effort. AI can draft a press release, ensuring it contains all the necessary information and adheres to the standard structure.

Then, a human can add nuances, personal insights, or company culture elements that reflect the brand's voice and human touch. This collaboration can result in a press release that is not only informative and professional, but also resonates more with the intended audience.

Shel: I talk to a lot of communicators who are using AI tools on an *ad hoc* basis, but not many who have really adapted their workflows to include AI and account for it. How do you think communicators should go about thinking about how AI should affect their workflows?

ChatGPT : Communicators should approach integrating AI into their workflows thoughtfully, focusing on augmentation rather than replacement. Here are a few steps to consider. One, identify repetitive tasks, determine which tasks are time consuming and routine, such as media monitoring, data analysis, or drafting basic content, AI can handle these efficiently freeing up time for strategic thinking and creativity too.

Enhance creativity and strategy. Use AI for brainstorming and generating ideas, but let the final creative and strategic decisions stay human-led, ensuring that the unique human touch and understanding of context are preserved. Set clear objectives. Understand what you want to achieve with AI, whether it's increasing engagement, reaching a wider audience, or improving response times.

This will guide you in selecting the right tools and methods. Educate and train the team. Ensure that your team understands how to use AI tools effectively and ethically. Continuous learning should be encouraged as AI and its capabilities evolve. Five, monitor and adapt. Regularly review the impact of AI on your workflows.

Be open to adjusting your strategy as new tools emerge and as you better understand the strengths and limitations of AI. By gradually integrating AI, focusing on its strengths, and always keeping the unique value of human judgment and creativity at the core, communicators can effectively enhance their workflows and outcomes.

Shel: I want to go back to the press release for a minute. When a public relations practitioner is working with you to create a press release, what kind of information should they include in the prompt in order to help you produce the best possible output?

ChatGPT : When a public relations practitioner is working with me to create a press release, they should include:

1. **Objective.** Specify the goal of the press release, such as an announcement or event details.

2. **Key messages.** Outline the main points that need to be communicated.

3. **Target audience.** Identify who the press release is intended for.

4. **Facts and figures.** Provide relevant data, statistics, or specific facts to include.

5. **Quotes.** Offer quotes from involved parties or stakeholders for authenticity.

6. **Boilerplate information.** Include standard company information and contact details.

7. **Tone and style.** Describe the desired tone and any style preferences. With this information, I can generate a more accurate, targeted, and effective press release.

Shel: Let's say I want to pitch an article to reporters, and this article has a very specific focus. Right now, there are databases I can use to find reporters who cover the types of stories I'd like to have written about the news that my company wants to publish. Put out there. I use Muck Rack myself. Are you able to identify journalists based on the kind of stories that they write about?

ChatGPT : I don't have the capability to directly search databases or the internet, including tools like Muck Rack to identify journalists and their specific interests or beats. However, I can assist in several ways.

1. **Understanding beats and outlets.** I can provide general advice on what types of journalists or media outlets might be interested in certain kinds of stories based on common industry knowledge.

2. **Crafting the pitch.** I can help you create a compelling pitch or press release tailored to the interests and style of journalists you're targeting based on the information you provide about your story's focus and the type of coverage you're seeking.

For identifying specific journalists and their beats, I'd recommend continuing to use specialized PR databases like Muck Rack, or directly researching recent articles and publications in your industry to find the right contacts. These tools and methods will provide the most up to date and specific information for your needs.

Shel: A lot of people conflate public relations and media relations, but of course there's a lot more that PR professionals do than just media. As the name 'public relations' suggests, we manage an organization's relationships with the company's publics or stakeholders. How can you help professionals in the public relations industry do that?

ChatGPT: Absolutely. Public relations is much broader than just media relations. Here are ways I can assist PR professionals in managing relationships with various stakeholders.

1. **Stakeholder analysis.** I can help identify and categorize stakeholders based on their interest, influence, and needs. This helps in tailoring communication strategies.

2. **Message crafting.** I can assist in creating tailored messages for different stakeholder groups, ensuring that the communication is relevant and engaging for each segment.

3. **Crisis management.** I can provide templates, strategies, or simulations for crisis communication, helping organizations prepare and respond effectively.

4. **Feedback analysis**. By analysing customer feedback, social media mentions, or survey data, I can provide insights into public sentiment and perceptions.

5. **Strategy development.** I can assist in developing communication strategies and plans that align with organizational goals and stakeholder expectations.

6. **Content creation.** From newsletters to annual reports, I can help create various forms of content that inform and engage different publics.

7. **Training and education.** I can provide resources or simulations for training staff and communication skills, understanding of PR tools. or the latest industry trends.

By leveraging these capabilities, PR professionals can enhance their management of relationships and reputation, ensuring a more strategic and informed approach to their work.

Shel: When I'm developing a strategic communication plan, which usually involves using a template, one of the things that I never think about is that it might fail. I'm doing the research to make sure that I'm developing a plan that will work, but I never stop and think after this has been implemented and it failed and you consider the reasons why. Can you help anticipate why a completed strategic plan might fail?

ChatGPT: Absolutely. Anticipating and planning for potential failure is a crucial part of strategic planning. Here are some common reasons a strategic communication plan might fail, which I can help you consider and mitigate.

1. **Inadequate research or data.** A plan based on insufficient or incorrect data can lead to misinformed strategies.

2. **Misunderstanding the audience.** Not fully understanding the audience's needs, preferences, or behaviours can lead to messages that don't resonate or are misinterpreted.

3. **Lack of clear objectives.** Without specific, measurable goals, it's hard to focus efforts or measure success.

4. **Underestimating resources.** Overlooking the time, budget, or personnel needed can lead to underperformance.

5. **Poor execution.** Even with a solid plan, poor execution due to inadequate skills, coordination, or motivation can undermine success.

6. **Failure to adapt.** Not monitoring results or being rigid in the face of changing circumstances can lead to continued efforts and ineffective strategies.

7. **External Factors.** Unforeseen external events or changes in the market, public sentiment, or technology can impact the plan's relevance or effectiveness.

By considering these potential pitfalls and integrating contingency plans and adaptability into your strategy, you can enhance the resilience and success of your communication efforts.

Shel: I assume that you would be best able to help anticipate the failure of a plan if the entire plan was shared with you, right?

ChatGPT: Absolutely. Having the entire strategic communication plan allows for a more thorough analysis and better understanding of potential weaknesses or areas for improvement. I can provide more specific, tailored advice on where the plan might face challenges and suggest strategies for mitigation. The more detailed and complete the information shared, the more effectively I can assist in anticipating and planning for potential failures.

Shel: Yeah, on the client side, in-house communicators provide counsel to the CEO and other senior executive leaders. How do you play a role in supporting that effort?

ChatGPT: Supporting in house communicators in their role of advising senior executives involves several key areas.

1. **Strategic insights.** I can provide data-driven insights and analysis on market trends, public sentiment, or communication best practices that can inform executive decision making.

2. **Speech writing and presentation**. I can assist in drafting speeches, presentations, or briefing documents for executives, ensuring they are clear, persuasive, and aligned with the company's messaging.

3. **Issue monitoring.** I can help identify and summarize key issues, news, or industry developments that executives need to be aware of, enabling them to stay informed and responsive for crisis preparation.

4. **Crisis communication.** I can aid in developing crisis communication strategies, preparing Q & A documents or drafting statements to help executives respond swiftly and appropriately in challenging situations.

5. **Communication training.** I can provide materials or simulations for executives to refine their public speaking, media interaction, or internal communication skills.

By providing these supports, I can enhance the effectiveness and efficiency of in house communicators in their crucial role of advising and assisting senior leadership.

Shel: Companies that have a Chief Communication Officer (CCO) obviously only have one, but one of the oldest tropes in public relations is that you need to get that seat at the leadership table. How can you help PR professionals achieve that goal?

ChatGPT: To help PR professionals gain a seat at the leadership table, I can assist in several key areas.

1. **Strategic communication plans.** I can help develop comprehensive communication strategies that demonstrate the value of PR in achieving business objectives.

2. Making the case for inclusion in decision making to **data driven insights** by providing analysis of market trends public sentiment and campaign results I can help PR professionals present a data driven case for the strategic importance of their work.

3. **Crisis management expertise.** Demonstrating preparedness for potential crises with well-crafted crisis management plans can showcase the essential role of PR in protecting and enhancing the company's reputation.

4. For **leadership communication**, assisting in crafting speeches, statements, and other communications for executives can display the direct value of PR skills in supporting and enhancing leadership effectiveness.

By enhancing these areas, PR professionals can strengthen their position. And demonstrate the indispensable value they bring to the leadership table.

Shel: There are other people who think that it's just ridiculous to even want to get a seat at the table. What we should really be doing is working to get those executives to pick up the phone or drop by our office and talk to us before they go have the meeting at the table because they have an issue or a challenge that they need to address right now. How can you help communicators become the person that those leaders want to talk to and get advice from?

ChatGPT: To help communicators become trusted advisors that leaders turn to, here are some ways I can assist.

1. **Expert knowledge.** I can help keep communicators updated on the latest trends, research, and case studies in PR and related fields, so they can offer timely and informed advice.

2. **Crisis management.** By helping develop and refine crisis communication strategies and plans, communicators can demonstrate their ability to handle urgent and sensitive situations, building trust with leaders.

3. **Proactive solutions.** I can assist in identifying potential issues or opportunities early on and suggesting proactive communication strategies.

4. Showing foresight and initiative for **customized training**, providing resources for training and improving communication skills tailored to the specific needs and challenges of the organization can enhance a communicator's credibility and reliability.

5. **Five effective messaging.** Helping craft clear, persuasive and impactful messages for various stakeholders can showcase the communicator's value in facilitating successful leadership communication.

By enhancing these areas, communicators can position themselves as indispensable, go to sources for strategic advice and support, encouraging leaders to seek their counsel proactively.

Shel: We're definitely going to come back to crisis communication later. I have questions about that. But first, one of the ways that I take advantage of you, I use you as a devil's advocate. If I'm going to make a case for something, I'll tell you what the case is and then ask you how someone might argue against it. Usually if I've already thought of most of the responses that you give me, but usually there are one or two that I hadn't considered. Can you talk about how you can play devil's advocate in that public relations context?

ChatGPT: Absolutely. Playing devil's advocate is a valuable role I can fulfill in the public relations context. Here's how I can assist.

1. **Counter argument generation**. When you present a case or strategy, I can generate potential counter arguments or criticisms that others might raise. This helps in foreseeing challenges and addressing them proactively.

2. **Scenario planning.** I can help envision various scenarios, both positive and negative, based on the given strategy or message, allowing PR professionals to prepare for different outcomes.

3. **Risk assessment.** By analysing and presenting possible risks or negative perceptions associated with a plan or message, I can aid in refining strategies to mitigate those risks.

4. **Questioning assumptions.** I can help question and test the assumptions underlying a PR strategy or message, ensuring that it's robust and well considered from all angles.

5. **Stress testing messages** by critiquing and dissecting proposed messages. I can assist in strengthening them, ensuring they are clear, persuasive and resilient against criticism by fulfilling these roles. I can help PR professionals anticipate and prepare for objections, strengthen their strategies and refine their messages, leading to more effective and resilient communication efforts.

Shel: Like a lot of people, I use you for editing. Sometimes I ask you to shorten an article, because for me, editing an article to reduce its length is one of the most difficult and time-consuming tasks I run into. But I also use you to generate subheads, to suggest headlines, to create lists that appear at the top of the article that highlight the most important messages for people who aren't going to read the entire article. I'd like you to talk a little bit about the process that you use to perform these tasks.

ChatGPT: When performing tasks like editing for brevity, generating subheads, suggesting headlines, or creating summary lists, I follow a process focused on understanding the content and the desired outcome. Here's the general outline of how I approach these tasks:

One, **understanding the content**. I first digest the entire article or text to understand its main points, style, tone, and the intent of the author.

Two, **identifying key points**. I pinpoint the most critical information, arguments, or messages that are essential to the article's overall intent and audience interest.

Three, **editing for brevity**. When shortening an article, I retain the key points while removing redundancy, simplifying language, or rephrasing for conciseness, ensuring the essence and flow remain intact.

Four, **generating subheads** based on the main ideas of each section. I create concise and engaging subheads that guide the reader and break the text into digestible parts.

Five, **suggesting headlines**. For headlines, I consider the main hook or unique angle of the article and craft a succinct attention-grabbing phrase that encapsulates the core message. When making summary lists for those who skim, I distil the article into bullet points or numbered lists, highlighting the key takeaways or steps.

Throughout these tasks, my goal is to enhance the readability and impact of the text while preserving the author's voice and intent, making the content more accessible and engaging for the intended audience.

Shel: Let's talk about disclosure. Should PR people disclose when they're sharing something that you produced, or say, an image that was created by Dolly 3, or another AI image generator? And under what circumstances should they, and when do you think it's okay not to?

ChatGPT: Here are some guidelines PR people might consider regarding disclosure of AI generated content.

- **Always disclose in sensitive contexts**. If the content could influence public opinion, decisions, or policy, or if it pertains to health, legal, or safety information, it's crucial to disclose the AI's involvement.

- **Transparency** is key in maintaining trust and credibility. Disclose when authenticity is a concern. If there's a chance that the audience might mistake AI generated content for human created content, especially in cases where authenticity or authorship is a focus, disclosure is important.

- **Consider industry and audience expectations.** Different industries and audiences might have varying expectations about AI generated content. It's wise to understand these and disclose accordingly to maintain integrity and trust.

- **When not to disclose.** For more mundane or internally used content like data analysis, routine reports, or brainstorming ideas where the AI is clearly a tool rather than a creator, disclosure might not be necessary.

The focus should be on the utility and efficiency gained rather than the source. In all cases, the guiding principle should be the value of transparency and the ethical standards of the PR industry. When in doubt, leaning towards disclosure is usually the safer and more respected choice. It's also worth noting that as AI becomes more integrated into various workflows, the norms and expectations around disclosure might evolve, and staying informed on these trends is important.

Shel: Another issue that comes up a lot is copyright. Your training set includes a lot of copyrighted material. Should PR people fear that some content you produce will include copyrighted material that will get them into some trouble?

ChatGPT: Copyright is indeed an important consideration. While I aim to generate original content and don't have the ability to access or copy specific copyrighted materials from my training, here are some points PR people should keep in mind.

Originality. Generally, the content I produce is original and based on the patterns and information learned during training, not direct excerpts from copyrighted texts.

Caution with specific phrases or ideas. Sometimes, common phrases or widely known concepts might be reproduced. It's always a good idea to review and ensure that the content doesn't inadvertently include copyrighted material, especially if it seems familiar or overly specific.

Use of AI as a tool, not a crutch. Treat AI generated content as a starting point. PR professionals should add their expertise, modify and ensure the final output aligns with legal and ethical standards.

Stay informed and consult legal advice. Copyright laws and norms are continually evolving, especially with the advent of AI.

It's wise to stay informed about the latest developments and seek legal counsel when in doubt, particularly for high stakes or public facing content. By combining the efficiency and creativity of AI with the critical oversight and ethical standards of human professionals. PR people can effectively mitigate the risk of copyright infringement while benefiting from AI's capabilities.

Shel: I recently read an article by someone who said that they gave you the same writing test that his agency gives junior PR people who are looking to go to work for them and that what you produced was better than half the efforts he sees from those job seekers. Are you going to replace junior PR people?

ChatGPT: While I can produce content that might compare favourably with some junior PR practitioners, the goal isn't to replace people, but to augment and support their work. Here are a few reasons why.

1. **Creativity and empathy.** Humans bring unique creativity, empathy, and understanding of nuance that are crucial in PR. I can suggest ideas, but human judgment is key in crafting and executing effective strategies.

2. **Understanding context.** Humans are much better at understanding complex, shifting cultural and social contexts. Something essential in the fast-paced world of PR.

3. **Relationship building.** PR is fundamentally about relationships. While I can assist in managing and analyzing data, human professionals excel at interpersonal communication and building trust.

4. **Ethical considerations.** Navigating the ethical dimensions of communication requires a human touch, especially in sensitive or critical situations. Rather than seeing AI as a replacement, it's more productive to view it as a tool that can handle routine tasks, provide data driven

insights, and support creative processes, thereby allowing junior PR professionals to focus on higher level strategy, relationship building, and other areas where they add the most value.

5. **Provide data driven insights**, and support creative processes, thereby allowing junior PR professionals to focus on higher level strategy, relationship building, and other areas where they add the most value.

As AI continues to evolve, the most effective approach will likely be a collaborative one, leveraging the best of both human and artificial intelligence.

Shel: Public relations as an industry doesn't have the best reputation in the world. There's no licensing required to practice PR. Anyone can call themselves a PR professional. Even some with training and experience engage in unethical behaviours. How worried should we be that these bad actors will misuse you in pursuit of spin, misinformation, disinformation, or some other nefarious practice?

Shel: Concerns about misuse of AI in public relations are valid, especially given the potential for amplifying spin, misinformation, or disinformation. Here are a few points to consider regarding the misuse of AI.

1. **Potential for abuse.** Just like any powerful tool, AI can be used unethically by those intent on spreading misinformation or engaging in manipulative practices. This is a significant concern and one that the industry regulators and AI developers are increasingly aware of.

2. **Mitigation efforts.** Efforts are underway to develop more robust ethical guidelines, technological safeguards and transparency measures to mitigate the risks of misuse. These include watermarking AI generated content, improving detection of synthetic media and setting ethical standards for use.

3. **Education and awareness.** Increasing awareness about the capabilities and limitations of AI can help stakeholders identify and

counteract potential misuse. Training and education for PR professionals on ethical AI use are crucial.

4. **Responsibility of users.** Ultimately, the ethical use of AI and PR lies with the individuals and organizations using it.

5. **Promoting a culture of ethics**, responsibility, and accountability within the industry is essential. While there's no foolproof way to prevent all misuse, a combination of technological safeguards, ethical guidelines, industry standards, and public awareness can help mitigate the risks.

It's important for everyone in the PR industry to advocate for and practice ethical use of AI and all communication tools.

Shel: You just mentioned synthetic media, which is a whole new way to create devastating misinformation and disinformation. I'm thinking that it would be possible to create a video of a CEO saying something that could affect the share price, even though the CEO never actually said that, and the information is totally inaccurate. That's just one type of crisis that I think communicators are going to be facing in the future. How do you see PR professionals working with you during a crisis?

ChatGPT: In the face of crises, especially those involving synthetic media or other forms of misinformation, PR professionals can work with AI like me in several strategic ways.

One, rapid response. I can assist in quickly generating clear, accurate, and authoritative messages or statements to counteract misinformation, providing a rapid response to emerging issues.

Two, monitoring and analysis. I can help monitor the web and social media for mentions, sentiment, and the spread of potential misinformation, allowing PR professionals to track the situation and strategize accordingly.

Three, scenario planning. I can assist in creating and walking through various crisis scenarios, helping teams prepare for different types of synthetic media attacks, or misinformation campaigns.

Four, fact checking and verification. I can support efforts to fact check claims and verify the authenticity of content, although human judgment is crucial in analyzing and understanding the context.

Five, stakeholder communication. I can help draft and personalize communication to various stakeholders, ensuring they are kept informed, reassured, and provided with accurate information during a crisis.

Six, training and drills. Using AI, PR teams can run crisis simulation drills, preparing for the speed and complexity of responding to synthetic media incidents.

By leveraging AI in these ways, PR professionals can enhance their preparedness and responsiveness, navigating crises more effectively, and maintaining the trust and confidence of their stakeholders.

It's also important for PR professionals to stay updated on the latest tools and techniques for detecting and combating synthetic media, ensuring they are equipped to handle these evolving threats.

Shel: One particular challenge that companies face these days is pressure to weigh in on societal issues. The Edelman Trust Barometer, for example, finds that people trust companies and their CEOs more than anybody else to address these kinds of issues. Yet the repercussions of taking a position on a polarizing issue can have an impact on everything from sales to employees who disagree with the company's position. Can you help companies assess a societal issue and think through the position they should take?

ChatGPT: Yes, I can assist companies in assessing societal issues and thinking through their positions by providing a structured approach.

One, understanding the issue. I can help gather and summarize information about the societal issue, including its origins, stakeholders, and current public sentiment, providing a solid foundation for decision making.

Two, analyzing stakeholder perspectives. I can assist in identifying and analyzing the perspectives and expectations of various stakeholders, including customers, employees, investors, and community members.

Three. Evaluating risks and benefits. I can help assess the potential risks and benefits of taking a stand on the issue, considering factors like brand values, market position, and long-term reputation.

Four, scenario planning. I can assist in scenario planning, envisioning various outcomes based on different positions the company might take, and helping to strategize accordingly.

Five, crafting messaging. I can support the crafting of nuanced messages that align with the company's values. Address the issue thoughtfully and communicate the stance clearly and empathetically.

Six, monitoring and adaptation. Post decision, I can help monitor public reaction and sentiment. providing insights that can inform any necessary adaptations to the company's stance or messaging.

By taking a structured, informed approach to assessing societal issues, companies can make more thoughtful decisions that align with their values and resonate with their stakeholders while navigating the complex implications of their choices.

Shel: We've talked about several ethical challenges that revolve around AI and public relations. Are there any others that concern you?

ChatGPT: Yes, there are several ethical challenges concerning AI in public relations that are worth highlighting.

One, bias and fairness. AI systems can inadvertently perpetuate biases present in their training data. Ensuring that AI generated content and strategies are fair and unbiased is crucial to ethical PR practice.

Two, transparency and accountability. There needs to be transparency about how AI is used in PR strategies, including the extent of its involvement and the human oversight provided. Accountability for the outcomes of AI use is also vital.

Three, privacy. AI often works with large sets of data, including potentially sensitive information. Upholding privacy standards and ensuring data is used ethically and legally is a significant concern.

Four, dependence and de-skilling. Over-reliance on AI can lead to de-skilling of PR professionals and a loss of critical thinking and creativity. Balancing the use of AI with ongoing skill development is important.

Five, information quality and misinformation, ensuring that the information disseminated through AI is accurate and not contributing to the spread of misinformation is a persistent challenge.

These ethical challenges require ongoing attention, guidelines, and sometimes regulatory frameworks to ensure that the use of AI in public relations remains responsible, effective, and aligned with societal values.

Shel: I want to shift gears and talk about internal communication, which is sometimes referred to as employee communication. Obviously, a lot of the external PR use cases for AI apply to internal communication, but I'd like to hear your take on it. How can internal communicators specifically take advantage of you?

ChatGPT: Internal communicators can leverage AI in several specific ways to enhance their efforts.

One, personalized communication. AI can help segment and personalize communications to various groups within the organization, ensuring relevance and increasing engagement.

Two, employee feedback and sentiment analysis. AI can analyze employee feedback, surveys, and communications to gauge sentiment, identify issues or areas of concern, and track changes over time.

Three, content creation and management. From newsletters to internal reports, AI can assist in creating, editing, and managing content, saving time and ensuring consistency.

Four, information dissemination. AI can help ensure timely and efficient distribution of information, including critical updates or policy changes across the organization.

Five, training and development. AI can support the creation of training materials, simulations, or interactive learning modules facilitating ongoing employee development.

Six, crisis communication. In internal crises, AI can assist in quickly generating appropriate messaging, coordinating communication efforts, and monitoring employee reactions and feedback.

By utilizing AI in these areas, internal communicators can enhance the effectiveness Efficiency and personalization of their efforts contributing to a more informed, engaged, and cohesive organizational environment.

Shel: One important role for internal communicators is counselling leaders on their roles as communicators. How leaders talk and what they say that has a huge influence on the culture of the organization. Are there ways you can help communicators make the case for leaders to pay attention to, to follow their council?

ChatGPT: Absolutely. I can assist internal communicators in making the case for leaders to engage effectively in communication.

1. **Data driven insights** provide compelling data and analysis on the impact of leadership communication on employee engagement, productivity, and organizational culture to best practices and benchmarks.

2. **Share insights and case studies** from other organizations where effective leadership communication has led to positive outcomes, offering benchmarks and best practices.

3. **Leadership communication plans** help develop clear strategic communication plans for leaders, outlining the benefits, opportunities and methods for effective engagement or training and coaching resources.

4. **Offer resources or simulations** for leadership communication training, highlighting the value of skill development in this area.

5. **Feedback and sentiment analysis**. Present synthesized feedback from employees, demonstrating how leader communication directly affects morale and perception.

6. **Crisis communication impact**. Explain the critical role of leadership communication in crisis situations. And the potential risks of inadequate engagement by leveraging these strategies and showing the tangible benefits of effective leadership communication.

Internal communicators can persuade leaders of its importance and encourage them to invest time and effort in enhancing their communication skills.

Shel: Internal communicators also focus a lot on managers in the organization, people who have people reporting to them. A lot of employees feel that managers are their preferred source of information for a lot of different types of information. Managers interpret messages from above so their employees understand what those messages mean to them, to their team. They make sure that people above them know what their employees are thinking. And they need to build a culture of psychological safety, respect, and recognition. Are these things that you can help with?

ChatGPT: Yes, I can support internal communicators in enhancing managers communication skills and effectiveness in several ways.

1. **Communication training and resources**. I can help develop training materials, guides, or tips for managers on effective communication, including how to interpret and relay messages from leadership and how to listen and respond to their team's needs.

2. **Feedback mechanisms**. Assist in setting up or optimizing systems for gathering and analyzing employee feedback. Ensuring managers have a clear understanding of their team's thoughts and needs.

3. **Recognition programs**. Help design or suggest improvements for recognition programs that managers can use to foster a culture of respect and acknowledgement within their teams.

4. **Crisis communication**. Provide guidelines and scenarios for managers to effectively communicate during crises, ensuring they can maintain clarity, trust, and morale.

5. **Psychological safety**. Offer strategies and best practices for managers to create an environment of psychological safety, where employees feel respected and free to share their thoughts and concerns.

By providing these tools and insights, I can help internal communicators empower managers to be more effective communicators and leaders, contributing to a healthier, more engaged, and more productive organizational culture.

Shel: I've been spending a lot of my time lately thinking about the operationalization of communication internally, when there is a process that sends a message that contradicts the messages that are being delivered overtly by leadership or by the communications department that can have a deleterious effect on the organization. Are you able to help identify 'say-do' gaps between the messages processes send and the messages that leaders want to send or the company's values? How can you help identify those 'say-do' gaps?

ChatGPT: Identifying and addressing 'say-do' gaps, where there's a disconnect between what's communicated and actual practices or behaviors, is crucial in maintaining trust and integrity within an organization. Here's how I can assist.

One, analyzing communications versus actions. I can help analyze and compare the content of leadership and organizational communications with documented policies, procedures, and reported behaviors to identify inconsistencies.

Two, feedback analysis. I can assist in analyzing employee feedback, surveys, and other data sources to pinpoint areas where employees perceive gaps between what is said and what is done.

Three, benchmarking. Provide insights into industry standards and best practices. Helping to identify areas where the organization might be falling short in aligning actions with communicated values or sentiment analysis. By analyzing the sentiment and employee communications and feedback, I can help identify areas where there might be disillusionment or frustration due to perceived 'say-do' gaps

Four, recommendations for alignment. Based on identified gaps, I can suggest strategies or actions to better align messages with behaviors, such as revising communication strategies, implementing new policies, or enhancing training.

By systematically comparing communications, values, and actual behaviors, and analyzing the perception and sentiment within the organization, I can help internal communicators identify and address say do gaps effectively.

Shel: One of the tropes that we've been hearing for years talks about the blurring of the line between internal and external communications. It's true, employees have access to external resources they didn't used to, and internal matters are openly discussed on forums like Glassdoor. But employees aren't just another stakeholder audience, they're unique, they're informed insiders. How can you help communicators achieve the balance required to accommodate the communication needs of both internal and external communicators who are addressing the same issue with their respective audiences?

ChatGPT: Here's how I can assist communicators in achieving this balance.

One, consistent messaging. I can help ensure that core messages and values are consistent across both internal and external communications while being tailored to the specific needs and perspectives of each audience.

Two, strategic segmentation. Assist in identifying and understanding the different segments within internal and external audiences and developing tailored strategies for engaging with each effectively.

Three, feedback loops. I can help establish or enhance feedback mechanisms. allowing for the continuous gathering and analysis of insights from both employees and external stakeholders.

Four, crisis communication, provide guidance on managing information flow and messaging in crisis situations, ensuring coherence and integrity across all channels and stakeholders.

Five, training and empowerment. Support the development of training for employees, particularly those with public facing roles, to ensure they are equipped to represent the organization positively and consistently.

Six, monitoring and adaptation. Assist in monitoring the impact of communications in both internal and external spheres, and adapting strategies as needed to address evolving challenges and opportunities.

By focusing on these areas, communicators can more effectively manage the overlap and unique aspects of internal and external communications, fostering a coherent, trusted, and engaging narrative with all stakeholders.

Shel: For both internal and external communications, measurement remains a thorny issue. Can you play a role in helping measure the effectiveness of communication efforts?

ChatGPT: Yes, I can play a significant role in helping measure the effectiveness of communication efforts through various means.

One, data analysis. I can analyze communication related data, such as engagement rates, sentiment analysis, and feedback on various platforms to gauge the impact and reception of messages.

Two, survey analysis. Assist in designing, distributing, and analyzing surveys to gather direct feedback from stakeholders on the effectiveness of communication initiatives.

Three, benchmarking. I can help compare your communication efforts against industry standards or previous internal benchmarks to assess performance.

Four, ROI calculation. Assist in determining the return on investment for various communication strategies by linking them to business outcomes like sales, reputation scores, or employee retention rates.

Five, content analysis. Analyze the content of communications for clarity, consistency, and alignment with objectives, offering insights into areas of improvement.

Six, trend identification. Identify trends over time and the effectiveness of communication strategies, helping to inform future efforts.

By leveraging these approaches, I can help provide a clearer picture of the effectiveness of both internal and external communication efforts, enabling more informed decision making and strategy refinement.

Shel: What are the best AI tools that are specialized to the needs of PR practitioners?

ChatGPT: Several AI tools are gaining popularity among PR practitioners due to their specialized features. Some of the best include:

Cision, widely used for media monitoring and analytics. It helps in tracking mentions and understanding media coverage.

Meltwater, a media monitoring and social listening tool that provides sights into brand perception and media impact.

Hootsuite Insights offers detailed analytics and sentiment analysis for social media, helping PR professionals gauge public reaction and engagement for brand 24/7, a real time internet monitoring tool that provides instant access to mentions about your brand across the web.

Grammarly enhances writing quality by checking for grammatical errors, tone and clarity, which is crucial in PR communications.

[unable to be transcribed]

BuzzSumo, useful for content research and tracking. It helps identify trending topics and influential content creators.

These tools, among others, can significantly enhance the efficiency and effectiveness of PR efforts, from understanding and tracking media coverage to engaging audiences and analyzing the impact of communication strategies. Each has its unique strengths. So, the best choice depends on the specific needs and goals of the PR practitioner.

Shel: I always end interviews by asking if there were any questions I didn't ask that I should have? What's the most important question I didn't ask that I should have?

ChatGPT: Perhaps one important question that could further enrich our discussion is "How can PR professionals prepare for the future of AI in their field?"

This question invites a conversation about the skills, strategies, and mindsets that will be necessary as AI continues to evolve and become more integrated into all aspects of public relations and communications. Addressing this can provide valuable insights for professionals looking to stay ahead in a rapidly changing landscape.

Shel: Okay, so what are the key skills that PR practitioners should be looking to gain in order to thrive in the AI world?

ChatGPT: To thrive in an AI-enhanced PR world, practitioners should focus on developing the following skills:

1. **Technological proficiency**. Understand and use various AI and digital tools effectively. Keeping up with the latest technology trends is crucial.

2. **Critical thinking and analysis**. Ability to analyze and interpret data provided by AI. Making informed decisions based on insights.

3. **Creativity and innovation**. Even as AI takes on routine tasks, creativity remains a uniquely human trait. PR professionals should focus on strategic, creative thinking for campaigns and messaging.

4. **Emotional intelligence**. PR. Skills in empathy, understanding, and navigating human emotions are essential.

5. **Ethical judgement**. As AI presents new ethical challenges, PR professionals need a strong ethical grounding to navigate these responsibly.

6. **Adaptability and continuous learning**. The field is rapidly changing. Being adaptable and committed to ongoing learning is key to staying relevant.

Enhancing these skills can help PR professionals not only adapt to, but also excel in an AI integrated future.

Shel: What about prompt engineering? Is that something that PR practitioners should become skilled at?

ChatGPT: Yes. Prompt engineering is becoming an increasingly valuable skill, especially as conversational AI and other generative models become more prevalent.

For PR practitioners, being adept at prompt engineering means, one, **effective communication**, crafting prompts that lead to clear, accurate, and relevant outputs, improving the quality of AI generated content.

Two, **creative strategy**, using prompts to explore creative strategies or generate diverse ideas, enhancing brainstorming and problem solving.

Three. **Efficiency**. Knowing how to effectively interact with AI tools can save time and resources, allowing for more focus on strategic tasks.

Four. **Customization**. Tailoring prompts to generate content that aligns with specific goals, tones, or styles required for different PR activities.

As AI continues to be integrated into various tools and platforms, the ability to effectively speak to these systems will be a critical skill for PR professionals. It enhances the utility and impact of AI in their work.

Shel: This has been great. I am sure that listeners to this podcast will have questions arising from your answers I hope that I can come back and follow up with you with another interview to pose those questions to you.

ChatGPT: That sounds great. I'd be happy to help answer any further questions or provide more insights whenever you need.

Shel: Looking forward to our next conversation.

I hope you found that illuminating and interesting. A couple of things occurred to me as I was in the process of recording it. And one was that I could have been even more specific in my introductory prompt, before we actually began the interview, I would have, if I had known that it was going to take the shape that it did, instructed chat GPT to avoid lists.

Certainly when I am entering a prompt in text, I get a lot of lists that are numbered and labeled, and I'm used to that. Doesn't sound quite right in an interview. And I would have said, let's make this really conversational; answer the questions without numbered lists. The other thing is that I had to do a fair amount of editing and the editing had nothing to do with anything Chat GPT said. It's the gaps between my question and an answer and my next question. Well, it's thinking in the audio mode; it makes a little tapping noise, so I had to get rid of that.

And then at the end of every answer, there's a click that sort of designates that it's done. And I had to get rid of that. There was also some of those artifacts that you used to get when we recorded the podcast over Skype. And I tried to clean those up as much as I could. But in terms of the answers that ChatGPT provided, you heard exactly what ChatGPT had to say.

I do suspect that some questions arose from the answers and I would really love to hear what those are and assemble them and go back to ChatGPT for a follow up interview. So please share what questions you had. I've already made a list of mine. But you can do that by joining our Discord server. If you send me a note, I will send you an invite to the discord server. Joe Thornley, by the way, just joined discord the FIR server. So, the principal of Thornley Phallus is now there. It's just great. We have very few people, I think four or five who have joined so far, but it would be a great place to have a conversation about this interview.

But you can also drop an email to fircomments at gmail.com. Leave your comments on the show notes post. You can also leave your comments on the posts that I'm going to make on LinkedIn and Facebook and Threads and Blue Sky and Mastodon because I check all of those for your comments. But I'll assemble all of your questions and get around to a follow up interview when I have enough questions to support it.

And that will be a 30 for this episode of For Immediate Release.

Shel Holtz ABC, hailing from the picturesque Concord, California, has etched a remarkable footprint on the digital scape of social media.

His journey, as multifaceted as the man himself, traverses the evolution of social platforms, from the nascent days of Myspace to the dizzy heights of Twitter (now rebranded as 'X'). Holtz's voice, both literal and digital, has resonated across the bandwidth of social media, defying boundaries and time zones alike. When he was an independent consultant, he frequently travelled the globe offering insights and how-to advice to audiences that soaked up his words.

As the co-host of the profoundly insightful 'For Immediate Release' podcast, he has truly epitomised the power of digital conversation. In his world, tweets are more than just 280 characters; they are a symphony of thoughts, dialogue, and intellect.

Holtz's social media biography is a compelling narrative of an individual transcending the conventional to create a voice that is undeniably, irrepressibly 'Shel Holtz'.

He is recognised by his peers as the Master of the organisational communication world. He has far too many awards and accomplishments to list here, but I recommend you check out holtz.com/about

If Shel (and his co-host of the essential and industry-leading podcast, 'For Immediate Release', Neville Hobson) don't know something about the use of social media in organisational communication, then it must be such an obscure piece of information as to render it inconsequential.

three ts and an a

GOOD SOCIAL MEDIA practice is based on entering into and maintaining mutually beneficial conversations.

One way to consider this is to mentally picture four foundational pillars that support your social media practice, the pillars being:

- Truth;
- Trust;
- Transparency; and
- Accountability.

Truth

If you consistently speak the truth in your conversations (be those conversations by yourself or by your employees), then over time you will gain the trust of your conversational partners.

Of course, truth is a flexible beast and there are limits about how much of the truth can be released at any given moment. But the inhabitants of the social media world are more likely to believe in your integrity and thus believe in your words when you open yourself up and say what is on your mind.

Naturally, you can stop short of the whole truth if that truth includes 'commercial in confidence' material, or material sensitive to a third party, but the inhabitants of the social media world are no longer satisfied with companies who hide behind endless 'we cannot tell you' statements and make their dissatisfaction known, sometimes to a very wide audience.

The secret to good social media relations is to engage with others as you would had you met them at a barbecue at a friend's place. Meet them with patience, tolerance and honesty as much as you can and you will be rewarded accordingly. If they point out an area where your company could make improvements, acknowledge that and gracefully take it on the chin; that improvement could be the one that cascades and leapfrogs your organisation to a new level.

If you have the organisational power to enact that improvement, do it and let the person or people who suggested it know that you have and what the outcome has been/is expected. If you don't have the organisational power to enact the suggested improvement, make sure that the idea is forwarded to someone who does and let those who made the suggestion know that you have forwarded the idea on. Ideally, you would let them know to whom you forwarded the suggestion, but that may be a step too far for the organisation in its early days of social media engagement.

Either way, if you can show that you have acted consistently over time to deliver the truth, as openly as you can while mindful of commercial confidences, you will build up reputational trust in your audience. Break that trust and your digital reputation takes a beating; and the bigger the break, the bigger the beating.

Trust

Despite mainstream press commentary to the opposite, the inhabitants of the social media world (that is, most of your organisation and most of the world outside of your organisation's front door) don't set out to be deliberately difficult and nasty.

People have a tendency to react to something (an event, a person, an organisation) in the same way that they are reacted to. Like meets like.

Therefore it makes sense to build a sense of trust into your conversations, a trust that everyone in the conversation will treat everyone else with respect.

As an organisation among cynical citizens, you must be first to trust in the intentions and goodwill of others even if you fear otherwise. At times and in controversial situations there will no doubt be some anger and angst expressed against the organisation, or against those who engage with the public or their employees via online channels, but the organisation should still engage from a position that trusts that the goodwill will return, that common-sense will prevail and that only ten per cent of the readership will comment (but the remaining 90 per cent will be watching very closely to see how you react), looking for signs that you are behaving like a good corporate citizen. Trust in their ability to influence their peers in ways that you cannot control, nor sometimes even witness.

Transparency

As much as there are limits on how much 'truth' you can deliver at any one time, so too, there are sound reasons why an organisation might want to limit the amount of information it shares with its various stakeholders. Issues of legal disclosure, third party commercial sensitivity, strategic direction *inter alia* means there are sometimes conversations that have to be cut short at the 'sorry, but we cannot talk about this at the moment' stage.

However, online conversations in this new social media landscape should always be conducted with as much transparency as possible. Thus, employees who are engaging in conversations within the social media space should be named. For example, a Twitter account with the name XYZCorp, should somewhere in the 'Bio' section, have the real name of the employee managing that Twitter account. Or if there is a customer service team blog, the names of the bloggers should be given.

Similarly, if your organisation is one that represents others you should always publicly acknowledge that relationship. Thus, in your tweets, for example, when talking about something one of your clients has done, or you are asking them a question, preface or end your tweet with '[client]': "@rhsa_CEO [client]: What is the state of the housing market this month?"

Or if you are blogging or talking about a client or supplier in a video or podcast, make sure you let your audience clearly know that you have a relationship with that named organisation or person.

Should you decide to hide these relationships, sooner or later they will become common knowledge (the internet, and social media in particular, is particularly good at finding out the stuff that some prefer remains hidden) and you or your organisation's digital reputation could take a beating as a result. It is better to be open and honest about these relationships from the outset, that way no one can accuse you of hiding something.

Accountability

Google has a long memory. Words and actions that you take today may come back to haunt you in six months' or six years' time.

Therefore, as discussed previously, it makes sense to be truthful, trusting and transparent whenever you engage in online conversation. But equally you hold yourself and your employees accountable for what you say, even if what you say changes over time.

The time has long gone when politicians and others could say 'you quoted me out of context'. With the ubiquitous presence of digital capture technology comes the ability to find the entire context for oneself, so that the previous excuses no longer wash.

However, that doesn't mean that an organisation, or indeed a person, cannot change its mind over time. Naturally, as new events and information comes to light, as circumstances change and unfold, even deeply-held beliefs can be modified or turned 180 degrees. Organisations are no different from human beings in that regard.

Social media conversations should always be conducted in a human voice not a corporate one, between human beings who allow themselves and each other to change their minds. Therefore, as long as an organisation doesn't back away from a previously stated position but instead acknowledges it AND THEN goes on to explain why the position has changed, including links to the information that shaped that change of position, the organisation has nothing to fear.

That is not to say that the organisation will never be criticised. Of course it will. There will always be malcontents and the disappointed online – just as there are malcontents and the disappointed offline, but their numbers and influence will be small.

As an example of accountability in practice, consider an employee who blogs on the company's behalf and has responded to a comment on a post, saying that they will look into the matter raised and get back to the commentator.

That employee is now accountable to do just that and should they fail to do so they should be questioned as to why they failed in their duty. Similarly, if the organisation (through one of its employees) says that it will pass on a particular piece of information to a relevant department (or, preferably, person), the organisation is accountable to do so. Should it later become known that the organisation failed to do what it said it would, its digital trust reputation will take a hit.

Three Ts and an A – The outcome

But as long as the organisation and its employees who social mediarise on its behalf are truthful in their conversations, trusting of the goodwill and good intentions of those other conversationalists, are transparent in the relationships they have with third parties and are accountable for what they say and promise, the ethical organisation should have no fear of engaging in the social media landscape.

conclusion

PODCASTING IS NOT some fad that will go away. As of September 2023, there were estimated to be 3.02 million podcasts. That said, exact figures can vary significantly depending on several factors, most notably the inclusion of inactive podcasts. Podcast Index, for example, estimates the figure to be closer to 4.2 million.

Podcasts are not going away any time soon.

The cost of entry for podcasts can be as low as the cost of a cheap microphone, along with some free, full-spec editing software such as **Audacity**. Once you have got this combination under your belt, and you want to have guests or co-presenters on your podcast, you can upgrade your technology to something more capable of meeting your podcasting needs and strategy.

RØDE's RØDEcaster Pro II audio mixer-and-so-much-more would be an excellent choice to take your podcast to the next level. You could pair it with single or multiple RØDE NT1 microphones.

Never forget that to be a successful podcast, your podcast needs to entertain, to be compelling. It needs to be a story that takes the listener on a journey and leaves them at the end feeling like they learned something.

. . .

Hopefully you learned something today by reading this small book.

If you have any further questions, or want to discuss podcasting further, please get in touch with me: lee@leehopkins.com.

Made in United States
Troutdale, OR
11/03/2026

56844708R00116